Owen Merriman

Gas Burners - Old and New

A Historical and Descriptive Treatise on the Progress of Invention in Gas

Lighting...

Owen Merriman

Gas Burners - Old and New
A Historical and Descriptive Treatise on the Progress of Invention in Gas Lighting...

ISBN/EAN: 9783337249779

Printed in Europe, USA, Canada, Australia, Japan

Cover: Foto ©berggeist007 / pixelio.de

More available books at **www.hansebooks.com**

GAS BURNERS

OLD AND NEW.

GAS BURNERS

OLD AND NEW.

A Historical and Descriptive Treatise

ON THE

PROGRESS OF INVENTION IN GAS LIGHTING;

EMBRACING AN ACCOUNT

OF THE

THEORY OF LUMINOUS COMBUSTION.

BY

"OWEN MERRIMAN."

Reprinted from the JOURNAL OF GAS LIGHTING.

London:

WALTER KING,

11, BOLT COURT, FLEET STREET, E.C.

1884.

W. KING AND BELL, PRINTERS,

12, GOUGH SQUARE, FLEET STREET,

LONDON.

PREFACE.

THE little work here presented to the public appeared originally in the pages of the *Journal of Gas Lighting*. In the hope that it may thereby become of service to a wider circle of readers, it has been revised and done into its present shape. The object of the writer will be attained if it is the means of lessening, in any degree, the suspicion and prejudice (born of ignorance) which, alas! yet prevail with regard to gas and gas lighting.

CONTENTS.

CHAPTER I.

INTRODUCTION.

The subject of gas-burners and the development of light from coal gas is of considerable interest, alike to the consumer and the producer of gas. When it is known that one burner may develop twice as much light as another, for the same consumption of gas—the first cost of the one being no higher than that of the other—its importance to the former will scarcely be disputed. To the gas consumer it is obviously of great value to know how he may most effectively and economically develop the illuminating power of the gas which is supplied to him; and so obtain the fullest return, in lighting effect, for the money which he expends. Not quite so obvious is its relation to the latter. To a person totally unacquainted with the recent history of gas lighting, and ignorant of the policy which has guided the most prosperous gas undertakings to their successful issues, it may appear that the manufacturer of gas is not closely concerned with the utilization of the commodity which he supplies. Such an one might argue, and with a certain show of reason, that the sole business of the gas maker is with its production; that after providing, in the consumer's service-pipe, a full and continuous supply of gas, of the stipulated quality, his care ends; and that henceforth the utilization and management of the illuminant rests with the consumer himself. But, by any one who is at all conversant with the subject, it will be readily conceded that the interest of the manufacturer of gas, in this matter, is only second to that of the consumer. In the gas industry, as in any other business undertaking, the concern prospers or declines according as the interests of the customers are considered or neglected. This has been conclusively demonstrated in the history of many gas undertakings. So long as their management was conducted in exclusive and selfish regard solely to their own internal affairs—looking with supreme indifference or careless apathy upon the needs of the consumers—so long was their career marked by difficulties and embarrassments. No sooner, however, were the claims of the consumers recognized, and efforts put forth to further their interests, than the prospects of the concern brightened; and by

Gas consumers and gas producers.

B

adhering to, and extending the same line of action, the goal of commercial prosperity was eventually reached.

Seeing, therefore, that the subject is of so supreme importance to consumers of gas, and that the interests of the consumer are closely interwoven with those of the manufacturer, it is eminently desirable that there should be more generally diffused a correct knowledge of the principles of economical gas consumption, and of the extent to which these principles are applied in the various burners which, from time to time, have been invented. No further apology ought therefore to be required in presenting to the reader the following disquisition on gas-burners. It may, however, be of advantage for me to state in brief, at the commencement, what are the objects I have in view, and what the chief considerations which have led me to write this treatise.

I purpose, then, to tell of the progress that has been made in apparatus for the development of light from coal gas; to relate how the crude and imperfect devices of the early inventors have been gradually improved upon; and, while not ignoring the drawbacks connected with recently invented burners, or the defects inherent to their construction, to show, in the superior achievements of these burners, how great an advance has been made upon the apparatus formerly in use. It will be, also, my endeavour to make plain the little-understood phenomenon of the production of light by the combustion of coal gas; and to show the extent to which the illuminating power developed is dependent upon the burner employed. That there is need for such information as I propose to furnish must be sufficiently obvious to any one who has considered the

Waste of gas. waste of gas which takes place through ignorance of the laws of its combustion, and through the use of defective burners. In a report presented to the Board of Trade by the London Gas Referees in 1871, it was stated that a number of burners had been tested, taken from various places of business in the Metropolis; the major portion of which gave out only one-half, and some of them not more than one-fourth, of the illuminating power capable of being developed from the gas. Although, since the time that report was penned, considerable progress has been made in the construction of burners, and in the more general adoption of efficient burners by the public, much yet remains to be done. Doubtless it would still be within the mark to assert that fully one-fifth of the gas consumed by the public might be saved by the adoption of better burners, and by the observance of the conditions necessary for their satisfactory operation; and when it is borne in mind that the gas-rental of the United Kingdom amounts to a sum of certainly not less than £9,000,000 per annum, the saving which might be effected assumes truly great proportions.

The field on which I propose to enter can hardly be said to be already occupied. Nowhere that I know of is the subject of gas-burners fully treated of in a manner available for the general reader. With the exception of the admirable chapter contributed by Mr. R. H. Patterson to "King's Treatise on Coal Gas," I am not aware that the subject has been dealt with to any complete extent by recent writers. But, admirable as is that contribution to the literature of the subject, being written for technical readers, it is neither so popular in style nor so elementary in character as to fulfil the purpose which I have in view in writing the present series of articles. Briefly stated, my sole purpose is to make the subject of the combustion of gas for the production of light intelligible to the simplest; and to present an interesting account of the progress of invention in the perfection of gas-burners. While passing lightly over many modifications of apparatus which have been of but limited or temporary service, I shall not scruple to dwell at length upon such burners as have done much to further the extension of gas lighting, or whose construction exhibits a considerable advance upon previous attainments. And while it will be my endeavour to clothe my remarks in such language as shall be "understanded of the people," in speaking of the theory of combustion I hope to be sufficiently explicit to enable my readers to form a clear conception of the scientific principles underlying the phenomena of which I treat.

A further justification—if such, indeed, were needed—for the appearance of this treatise might be found in the remarkable impetus which has been given, within recent years, to the perfection of the details of gas manufacture and the improvement of gas-burners. Of course, I refer to the beneficial consequences to the gas industry which have followed the brief, if conspicuous, career of electricity as an illuminating agent. That the interest in improved illumination which has been aroused by the short-lived popularity of the electric light, and the extravagant claims put forward on its behalf, have stimulated to the development of the resources of gas lighting, is sufficiently obvious to the most superficial observer. And not only has the manufacturer of gas been benefited, but the public have reaped no inconsiderable advantage. At the present day, gas is sold at a far cheaper rate, as well as of a higher quality, than at any former period. Nor is the advent of cheap gas the only direction in which the public have gained. Although not so patent to the majority, the improvements that have been effected in the methods of burning gas, so as to obtain the fullest advantage from its use, are calculated to confer benefits equally real, and not less valuable. It is hardly too much to say that the last few years have witnessed a greater advance in the

Progress of gas lighting.

apparatus employed in the combustion of gas than had been
effected during the whole previous history of gas lighting. This
being so, it may not be unacceptable if I attempt to pass in review
some of the various burners that have been invented and used for
obtaining light from coal gas; showing the successive improvements that are exhibited in their construction, and the extent to
which they apply the principles of combustion. It may be that
what I have to relate will awaken some minds to the consciousness
that gas lighting has not altogether retired into obscurity on the
advent of electricity—nay, that it has even assumed a bolder front;
and, with increased resources and accession of strength, is prepared
firmly to maintain its position as at once the most convenient,
economical, and reliable of artificial illuminants.

CHAPTER II.

FLAT-FLAME BURNERS.

THE FIRST GAS-BURNER.

The first gas-burner was a very simple and unpretentious contrivance. In one of the earliest works on gas lighting * we read: "The extremities of the pipes have small apertures, out of which the gas issues; and the streams of gas, being lighted at those apertures, burn with a clear and steady flame as long as the supply of gas continues." Familiar as it is to us, and from its familiarity unnoticed, the phenomenon presented by the flame thus produced continuing to burn "as long as the supply of gas continued," was doubtless, to the first experimenters, a wonderful sight. Though we may smile at the question, it is not difficult to understand the incredulity of the honourable member who, when Murdock was examined before a Committee of the House of Commons, in 1809, asked the witness: "Do you mean to tell us that it will be possible to have a light *without a wick?*" "Yes; I do indeed," replied Murdock. "Ah, my friend," replied the member, "you are trying to prove too much."

It was but natural, seeing that oil-lamps and candles were the only forms of artificial illumination in use prior to the introduction of gas lighting, that the earliest attempts at illumination by gas should be in imitation of the effects produced by those means. Accordingly we find that one of the first gas-burners employed was the Argand, modelled upon the oil-lamp of that name, which had been found to give superior results; while in more general use, and for some time almost the sole apparatus available, were single jets, giving a flame similar in appearance to that of a common candle, together with various combinations of these jets. A fair idea of the mode of illumination practised during the earliest period of gas lighting may be gleaned from the following extract from a paper describing the lighting of Messrs. Phillips and Lee's cotton-mill at Manchester, read before the Royal Society, in 1808, by Mr. William Murdock:— *The dawn of gas lighting.*

The gas-burners are of two kinds. The one is upon the principle of the Argand lamp, and resembles it in appearance; the other is a small curved

---

* Accum's "Treatise on Gas-Lights." Third edition, 1816.

tube with a conical end, having three circular apertures or perforations, of about 1-30th of an inch in diameter, one at the point of the cone, and two lateral ones, through which the gas issues, forming three divergent jets of flame, somewhat like a fleur-de-lis. The shape and general appearance of this tube has procured for it, among the workmen, the name of the "cockspur" burner.

Nor was much advance made upon these arrangements down to the year 1816, judging from Accum's "Treatise" (before cited), as

FIG. 1.—EARLY GAS-BURNERS.
(From Accum's "Treatise on Gas-Lights.")

the subjoined extract from that work, together with the above illustrations, will show :—

The burners are formed in various ways—either a tube ending with a simple orifice, at which the gas issues in a stream, and if once lighted will continue to burn with the most steady and regular light imaginable, as long as the gas is supplied; or two concentric tubes of brass or sheet iron are placed at a distance of a small fraction of an inch from each other, and closed at the bottom. The gas which enters between these cylinders, when lighted, forms an Argand lamp, which is supplied by an internal and external current of air in the usual manner. Or the two concentric tubes are closed at the top with a ring, having small perforations, out of which the gas can issue; thus forming small distinct streams of light.

It is interesting, in view of the present demand for increased illumination, and for burners of high illuminating power, to note the amount of light produced by the burners then in use. In Mr. Murdock's paper we find it stated that each of the Argands in use at Messrs. Phillips and Lee's establishment gave "a light equal to that of 4 candles (mould candles of 6 to the pound) ;" and each of the cockspurs "a light equal to $2\frac{1}{4}$ of the same candles." From which meagre results we conclude that, besides being burnt in an ignorant and wasteful manner, the gas consumed was wofully deficient in illuminating power.

THE BATSWING BURNER.

A notable advance was made when the batswing burner was invented. To whom we are indebted for this invention seems involved in some doubt. Although Clegg, in the historical introduction to his valuable work,* says, very distinctly, that "the batswing burner was introduced by a Mr. Stone, an intelligent workman employed by Mr. Winsor," it is not so much as mentioned by Accum, even in the third edition of his "Treatise;" and Accum, it may be remarked, was for some time closely associated with Winsor in the promotion of the latter's ambitious and visionary schemes. Yet, if Clegg's statement be correct, it would almost appear to fix the date of the introduction of this burner as prior to 1816. But to whomsoever is due the credit of its invention, certain is it that the batswing burner was a considerable improvement upon the old cockspur. Producing a better light for the gas consumed, it assisted to demonstrate still further the superiority of gas lighting over other methods of illumination; and as it could be supplied at a trifling cost, and contained no delicately adjusted nor easily injured parts, it enabled the benefits of the new method of lighting to be extended to wherever artificial light was required.

From the cockspur and single-jet burners the gas ascended in streams, rising into the air until it came in contact with sufficient oxygen to completely consume it. In order that this might take place without producing a flame of an inordinate length, and without much smoke, the orifices were restricted to a very small

Who invented the batswing burner?

Superiority of the batswing over the cockspur burner.

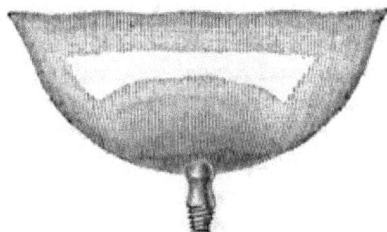

FIG. 2.—BATSWING BURNER.

size; and the gas issuing from these at considerable pressure tended to draw in, and mix with the air in its course. Besides the loss of illuminating power caused by this mixture of air with the

* Clegg's "Treatise on Coal Gas," 1841, p. 21.

gas flame (similar to what takes place in a Bunsen burner), the cooling influence upon the small body of flame of the mass of metal composing the burner, operated still further to reduce the quantity of light which the gas was calculated to yield. With the batswing the gas was spread out producing, when ignited, a thin sheet of flame, by which means the gas was enabled to combine more readily with the air necessary to effect complete combustion. The size of the flame being, in comparison with that of the cockspur, so much larger proportionately to the metal burner, the cooling effect of the latter was not so apparent. The increased size of flame, also, of itself, tended to improve the illuminating power; each portion of flame contributing to elevate and sustain the temperature of the whole, and so to heighten the intensity of incandescence to which the light-giving particles were raised.

Even with the Argands of that day, the batswing compared not unfavourably. The former burner, having the regulation of its *Batswing and* air supply under complete control, gives the best results when the *Argand* gas is supplied to it at a low pressure; as then the requisite *burners com-* quantity of air to ensure complete combustion of the gas can be *pared.* delicately adjusted by means of a chimney of suitable length. When the gas and air have been nicely adjusted to each other, the flame becomes extremely sensitive to any change of pressure in the gas supply; a diminution of the supply, by reducing the quantity of gas issuing from the burner without at the same time proportionately diminishing the supply of air, tends to destroy the illuminating power by the cooling action of the surplus air; while an increased pressure, by allowing more gas to issue than the air can consume, causes the flame to smoke. But at the time to which I now refer the principles of combustion were little understood, still less applied in the construction of burners. Besides this, the pressure of the gas in the mains was excessive; and there being no method adopted of controlling it at the burner, the construction of a good Argand was, under the circumstances, almost impossible. The batswing was not so prejudicially affected by an excess of pressure. Pressure to some extent was, indeed, required to enable the flame to attain its normal shape; while any excess forced the gas through the flame without permitting it to be raised to incandescence before being consumed, and although necessitating loss of light, caused no inconvenience like a smoking flame. Another important advantage which the batswing possessed over the Argand burner was its simplicity of construction; and the absence of accessories, such as the glass chimney—dispensing with the cleaning and attention which the latter required. Had the benefits of gas lighting been dependent upon the use of apparatus so fragile, and requiring so much care and attention as the Argand, the range of its applicability must have been considerably limited, and its prospects of commercial success much less assured. The introduction of a series of cheap but effective burners, however, altered the conditions of gas lighting, and marked the commencement of a new era in artificial illumination. The possibility of obtaining, by means of a burner so simple and apparently insignificant as the batswing, results little, if at all, inferior to what could be obtained by the use of the most complicated and expensive, was of advantage alike to the consumer and the producer of gas. To the former it gave the benefits of an increased illumination, without requiring any corresponding outlay; to the latter it promised a growing extension of the use of coal gas, and thus furnished the surest guarantee of future progress and prosperity.

THE UNION-JET, OR FISHTAIL BURNER.

The batswing had been for some years in extensive use before a burner was produced worthy in any degree to compare with it in respect to simplicity and efficiency. The invention of the union-jet, or fishtail burner, furnished a competitor equally simple; little, if at all, inferior as regards efficiency; and, to some extent, superior to the former burner in general adaptability. Although so much behind in point of time, the new burner speedily rivalled the older batswing in popular favour; and in its various modifications and improvements may be said, without fear of contradiction, to have received a wider application than any other gas-burner. As in the case of the batswing, so with regard to this burner: few details are recorded of its invention. But, slight as is the information available, such as we have is more satisfactory and more authentic than the meagre notice of Clegg, which is all that is known of the invention of the former burner. It appears to be established beyond doubt that the union-jet is the joint invention of Mr. James B. Neilson, the inventor of the hot-blast, and Mr. James Milne, of Glasgow, founder of the engineering firm of Milne and Son. About the year 1820, or soon after (as in that year Mr. Neilson was appointed Manager of the Glasgow Gas-Works), these gentlemen were experimenting with gas-burners, when they discovered that by allowing two jets of gas, of equal size, to impinge upon each other at a certain angle, a flat flame was produced, with increased light. This was the origin of the union-jet; so called from the manner in which the flame is produced. At first separate nipples were employed for the two jets; but, very soon, Mr. Milne hit upon the expedient of drilling two holes, at the required angle, in the same nipple. In this manner, with slight modifications, the burner has continued to be constructed down to the present day.

Who invented the union-jet burner?

The explanation of the preference accorded to this burner over its predecessor, the batswing, is to be found chiefly, I think, in the very different shapes of the respective flames produced by the two burners. The batswing, in its original form, produced a flame of great width, but of no corresponding height. The extremities of the flame, stretching out from the burner so far on either hand, were easily affected by an agitation of, or commotion in the surrounding atmosphere; a slight draught or current of air causing the flame to smoke at these points. The extreme width of flame also precluded the use of this burner in globes. The flame produced by the union-jet burner, as first constructed, was very different to the one just described. Longer than that of the batswing, and considerably narrower (but widening gradually from its base, at the burner, to its apex), it presented somewhat nearly the appear-

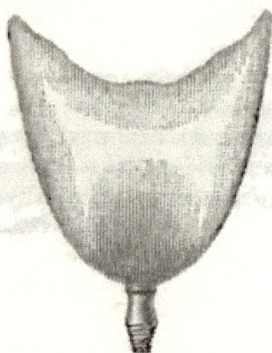

Fig. 3.—Fishtail Burner.

ance of an isosceles triangle; or more closely, perhaps (with its slightly-forked apex), the tail of a fish, from which resemblance it is commonly designated the fishtail burner. This form of flame was better adapted for use in globes, and also better withstood the effects of draughts. And it is perhaps not unreasonable to suppose that as in shape it approached more closely to the kind of flame with which the people had been familiar in oil lamps, the flame produced by the union-jet burner was more agreeable to the eye than that of the batswing, and that this seemingly trivial consideration will account, to some extent at least, for the undue favour shown towards it. For it must not be assumed, because of the widespread popularity to which the union-jet so early attained, and which it has continued to enjoy, that it was of necessity a better burner (in the sense of developing more light for the gas consumed) than the one which preceded it. On the contrary, in this regard it was not quite so effective as the batswing. Nor is this result surprising, looking at the different methods adopted in the two burners for producing the same effects of light and flame.

From the batswing burner the gas issued in a thin but widely-extending stream, presenting, when ignited, a continuous sheet of flame; its height and width depending upon the pressure at which the gas was supplied, but always offering an unbroken surface of flame to the air. Although, from the excessive pressures which, in the early days of gas lighting, were generally employed, the flame drew upon its surface too much air for the attainment of the fullest lighting efficiency obtainable from the gas; yet the form given to the issuing stream of gas precluded the air from entering the interior of the flame, and still further reducing its illuminating power. With the union-jet burner the conditions were greatly changed; and this latter evil, of the introduction of cold air into the interior of the flame, was one of the consequences entailed by the means it employed for producing its flame. From this burner the gas issued in two narrow streams, like single jets, which, directly after emerging from the burner, impinged upon each other at a given angle; the mutual shock given to the streams of gas when thus arrested causing them to spread out in a lateral direction, and (the high velocity at which the gas issued being expended) to unite, and ascend in a sluggish stream until consumed. That injury to the illuminating power of the flame should result from causes connected with the manner of producing it will be understood on considering some of the phenomena associated with the production of a gas flame.

Union-jet and batswing burners compared.

When a jet or stream of gas issues into a still atmosphere, it produces in its immediate neighbourhood, on all sides, an area of low pressure, to occupy which the contiguous air rushes in. Induced air currents are thus set up in close proximity to, and having the same direction as the issuing stream of gas, and varying in force with the pressure, or velocity, at which the gas issues. The non-luminous flame of the Bunsen burner, and of the so-called "atmospheric" burner employed in gas cooking and heating stoves (which is produced by burning a mixture of gas and air), is obtained by taking advantage of this tendency of a stream of gas, issuing under pressure, to draw air upon itself; and it is to the same circumstance that ordinary illuminating flames owe the continuous supply of air necessary to keep up combustion. For the effect is heightened when the gas is inflamed; because, the gaseous products of combustion being expanded by the intense heat to which they are subjected, their velocity of ascension is vastly increased. Having regard to these considerations, it will be clearly perceived how that, in producing the flame of the union-

How air is drawn upon a gas flame.

jet burner, the two streams of gas, in the act of combining together, drew into the very midst of the flame a portion of the air with which they were surrounded ; and this air, reducing the temperature of the flame, and diluting the illuminating gas by the inert nitrogen introduced, as well as by its oxygen causing a too early oxidation of the carbon particles in the flame, operated to reduce the illuminating power otherwise obtainable from the gas.

The foregoing remarks, it must be borne in mind, refer to the union-jet burner in its original form. Numerous improvements have been effected, from time to time, in its construction, as well as in that of the batswing, which, by reducing its liability thus to convey air into the flame, have increased its efficiency ; while, at the same time, the shape of the flame has been improved. Indeed, the result of successive improvements in the construction of both burners has been so to modify the shape of their respective flames that, in their latest and most improved form, the flames produced by the two burners are practically identical in appearance, although the manner of their production remains as widely diverse as at the first. The improvements that led up to, and the causes that produced this result, will be more fully explained in the sequel.

HOW LIGHT IS PRODUCED FROM COAL GAS.

I have before remarked that, in the early period of its use, one of the chief obstacles to the development of the lighting power of coal gas was the excessive pressure at which it was generally supplied. To understand the action of pressure in influencing the amount of light which a given quantity of gas will afford, it is necessary to know something of the nature and properties of flame. Moreover, the conditions upon which is dependent the illuminating power of a gas flame are so intimately related to each other, that the precise functions due to each cannot well be separated from the complete effect produced by the combined operation of all. I shall not, therefore, be needlessly digressing from my subject if, at this juncture, I explain the manner in which combustion takes place in the flame of an ordinary gas-burner. In doing this, I shall endeavour to clothe my remarks in very plain language ; using no more technicalities than are absolutely required by the exigencies of the subject. In this way I hope to make my meaning clear to the simplest. At the same time, without pretending to be scientifically complete, the explanation of the phenomena of combustion which I shall furnish will, I trust, be sufficiently explicit to enable the reader to form a right estimate of the principles which regulate the production of light when coal gas is consumed. The end chiefly kept in view is to show clearly the extent to which the degree of light evolved is dependent upon the burner employed, and the manner in which the gas is consumed. If my remarks are the means of causing the reader to look with intelligent interest upon the familiar phenomena of gaslight, they will not have been written altogether in vain.

Seeing that this treatise is compiled especially for those whose knowledge as to what coal gas consists of is extremely limited, it may be of advantage to preface my observations on its combustion, and the production of light therefrom, by a few remarks as to its composition. Coal gas, as generally supplied, is made up of a variety of distinct gases; of which, however, only some three or four exist in any considerable proportion. About 50 per cent., by volume (or half of the whole), is hydrogen; from 30 to 40 per cent. consists of marsh gas; while carbonic oxide is usually present to the extent of from 5 to 15 per cent. These three gases, which constitute the great bulk of what is known as common gas—that is, gas made from ordinary bituminous coal, as distinguished from that produced from the more costly cannel—are of little or no value as regards the amount of light they are capable of affording. The flames produced by the burning of the two former gases evolve much heat, but are of very feeble illuminating power. The latter gives a flame of a deep blue colour, producing scarcely any light, but, like the other two, an intense heat. The power of coal gas to yield a luminous flame is dependent upon the small quantity of heavy hydrocarbons which it contains—a constituent, or series of constituents, of which common gas only contains a proportion varying between 3 and 7 per cent., although in cannel gas it reaches as high as 15 or 20 per cent. These heavy hydrocarbons are gases composed, like marsh gas, of carbon and hydrogen; but containing in their composition, for each unit of volume, a greater aggregate of the two elements, as well as a relatively higher proportion of carbon, than exists in marsh gas. One of the simplest members of the series, and that which is usually present in by far the largest amount, is called olefiant gas. It contains twice as much carbon, combined with only the same quantity of hydrogen, as is contained in marsh gas. But besides olefiant gas there are minute quantities of other gases of the same series, having an analogous composition, but differing in the amount and relative proportions they contain of the two elements of which they are composed. All the gases of this series, when properly burnt, are capable of affording a brightly luminous flame; but when consumed alone it is somewhat difficult, on account of the high proportion of carbon which they contain, to effect their combustion without the production of smoke. It is, then, to the heavy hydrocarbons which are part of it—insignificant as their amount may appear—that the

What is coal gas?

luminosity of a gas flame is solely due. The other constituents which I have mentioned as forming so much larger a proportion of the whole, besides contributing to the heat of the flame, serve only to dilute these richer gases, and so promote their more complete combustion.

The various simple gases which constitute ordinary coal gas do not all burn together in the flame; the temperature required to effect their ignition being lower for some of them than for others. Thus, hydrogen is the first to burn, taking fire readily as soon as it issues from the burner; while the combustion of the heavy hydrocarbons does not commence until they enter the hotter portions of the flame, and is not completed until they reach its farthest extremity. Neither is the process of combustion in both cases the same. The former gas is at once completely consumed; the latter first undergo decomposition by the heat of the flame, being resolved into their elements—hydrogen and carbon—before being fully consumed. This decomposition of the hydrocarbons is a factor of supreme importance in the development of the lighting power of the flame. The hydrogen they contain, being more easily ignited than the carbon, burns first; and the latter is set free, in the solid form, as minute particles of soot. These particles of solid carbon, being liberated in the midst of the flame, are immediately subjected to its most intense heat; they thus become white hot before they reach the outer verge of the flame, and come in contact with sufficient oxygen to effect their complete combustion. The amount of light developed by any coal-gas flame is directly proportional to the degree of intensity to which the temperature of these carbon particles is raised, and the length of time they remain in the flame before being finally consumed. It becomes, therefore, a matter of considerable importance to know the conditions which are most conducive to the early liberation in the flame of free carbon, and the attainment by it of an exalted temperature.

Looking at the flame (say) of a common slit burner, it is seen to be divided into two sharply defined and wholly distinct portions. First, there is—immediately surrounding the burner head, and extending to some distance from it—a dark, transparent area, which, on closer examination, is found to consist of unignited gas enclosed in a thin envelope of bright blue flame. Second, there is (beyond this central area) a zone, or belt, of brightly luminous flame, white and opaque; the latter property indicating the presence of solid matter at this part of the flame. That the dark central portion of the flame consists chiefly of unignited gas may be shown in various ways, in addition to the evidence afforded by its complete transparency. Thus, if a small glass tube be taken, and its lower end

How gas
burns.

What is a
gas flame?

inserted in the flame at this point, the unburnt gas will pass up the tube, and may be lighted at its upper extremity. A splinter of wood

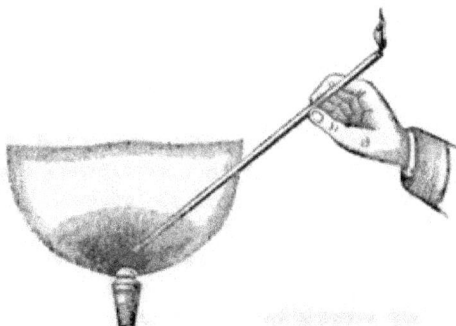

Fig. 4.—Showing the Two Zones of the Flame, and the Method of Demonstrating the Presence of Unburnt Gas in the Flame.

thrust through this portion of the flame is charred first at the two edges of the flame ; while, in like manner, a piece of platinum foil remains dull in the centre of the flame, and glows only at the points of contact with the outer air.   The presence of solid carbon in the luminous portion of the flame may be shown by inserting therein any cold substance (such as a piece of metal or porcelain), which, reducing the temperature of the heated particles of carbon below the point at which they are consumed, becomes instantly coated on its under surface with a deposit of soot.   Or, if the flame be suddenly cooled by gently blowing upon its surface, the same result is brought about ; clouds of soot are given off, and the flame "smokes." *

The existence, in the midst of the flame, of an area of uncon-sumed gas is due to the cold gas, as it issues from the burner, cooling the interior of the flame below the temperature required for its ignition, as well as to its not at once meeting with sufficient air for complete combustion.   The causes which affect the lumi-nous zone of the flame are not so readily explained.   It has been stated that the luminosity of the flame is due to the particles of carbon, which are separated out of the hydrocarbons in the gas,

---

* The behaviour of gas flames when exposed to the action of the wind (as exemplified in the naked lights of open markets and similar situations) affords an instructive illustration of the theory of luminous combustion. A sudden gust causes the flame to smoke, by reducing the temperature of the liberated carbon below the point at which it can combine with the oxygen of the air. A continuous wind blowing upon the flame destroys its luminosity altogether, because the heat-intensity of the flame is lowered below the temperature necessary to decompose the hydrocarbons ; consequently, these latter burn without the preliminary separation of carbon, and a non-luminous flame is produced—exactly as in the Bunsen or "atmospheric" burner.

being raised to a white heat. To decompose the hydrocarbons, a very high temperature is required; and, on account of the cooling effect of the stream of cold gas, this is not attained except at some distance from the burner. The abstraction of heat by the burner **How the** itself is also a cause of the reduction of the temperature of the **flame is cooled.** flame; and, on this account, burners of porcelain, steatite, or similar composition, being bad conductors of heat, have an advantage over those made of metal. So considerable is the cooling influence of the gas stream, that, within certain limits, the distance, from the burner head, at which the luminosity of a flame commences, is proportionate to the velocity with which the gas issues ; or, in other words, the pressure at which it is delivered from the burner. The effect is heightened by the tendency (which has been before remarked) of a stream of gas, issuing under pressure, to draw upon itself and mix with the surrounding air. Thus, with each increment of pressure the luminous zone of the flame is farther removed, until a point is reached at which the gas is so mixed with air before being consumed that the luminosity of the flame is completely destroyed.

But it must not be assumed, because of the foregoing remarks, that the pressure at which the gas issues from the burner is altogether an unmixed evil. In flat-flame burners it fulfils the im- **Effects of** portant function of promoting intensity of combustion, by bringing **pressure in** the white-hot particles of carbon into intimate and rapid contact **the gas supply.** with the air that is necessary for complete combustion. In Argand burners this duty is discharged by the glass chimney ; but with flat-flame burners it devolves entirely upon the pressure at which the gas issues from the burner. It will be seen, therefore, that the pressure of the gas is a factor of considerable importance in determining the amount of light afforded by a gas flame, as it is a matter requiring careful adjustment with each and every burner. On the one hand, with an excessive pressure the intensity of combustion is increased ; but the separated carbon does not remain so long in the flame. The area of luminosity is thereby decreased, and the total light yielded is reduced. On the other hand, with insufficient pressure the combustion is not energetic enough to raise the particles of carbon to a white heat ; consequently, the illuminating power of the flame is feeble, or else the carbon escapes unconsumed as smoke.

The thickness of the flame produced by any burner has also an important bearing upon the degree of light afforded ; and this property of thickness, again, is dependent upon the width of slit, in the case of batswings (or, in the case of union-jets, upon the size of orifices), and the pressure at which the gas is supplied. The thickness of the flame yielded by any burner will obviously vary inversely with the pressure at which the gas is supplied to it. With a thin flame, all parts of the flame are so completely exposed to the air, that the particles of carbon are no sooner raised to the temperature required to enable them to give out light than they are entirely consumed. With a thicker flame the carbon separated in the midst of the flame exists for a sensibly longer period of time in the white-hot state before it reaches the outside of the flame, and meets with sufficient oxygen for its complete combustion. Thus we find that the best flat-flame burners have comparatively wide orifices; while the pressure at which the gas is delivered from the burner is carefully reduced to the lowest point at which a firm flame is obtained, without smoke. Similarly, in the best Argands the pressure is considerably diminished within the burner, and the gas allowed to issue gently through relatively large holes; while the chimney is carefully adapted to draw upon the surface of the flame just sufficient air to completely consume the quantity of gas which the burner is calculated to deliver.

IMPROVEMENTS IN FLAT-FLAME BURNERS.

Although, there is no doubt, they were made empirically, and in ignorance of the real effects of pressure upon the flame, the first steps towards increasing the efficiency of flat-flame burners were in the right direction of reducing the excessive pressure at which the gas was formerly allowed to burn. They consisted in the adoption of simple arrangements for obstructing the passage of the gas through the burner, and so retarding its flow. The crudeness of the means which were employed is sufficient evidence that the end aimed at was, at best, but dimly discerned. The body of the burner was stuffed with wool, or pieces of wire gauze; which impeded the progress of the gas; reduced the quantity that would otherwise have been consumed; and, consequently, diminished the velocity with which it issued from the burner. Unfortunately, owing to the imperfect methods in use at that day for condensing and purifying the gas, the burners so constructed became choked with the tarry matters held in suspension, and carried forward by the gas; and so, after a comparatively short period of service, were rendered entirely inoperative. But, altogether apart from the inconvenience and loss thus entailed (which, when improved modes of manufacture had removed the cause, ceased to be experienced), the arrangement was ill adapted for the purpose which it was designed to serve. The rough and uneven nature of the material employed to stuff the burner caused the gas to eddy and swirl as it issued into the atmosphere, and prevented it being supplied equally to all parts of the flame. The consequence was

c

that the advantages which ought to have been derived from the diminished pressure were neutralized by the unsteady flow acquired by the stream of gas ; and the illuminating power developed by the flame was little improvement upon what could previously be obtained by the manipulation of the tap controlling the supply of gas to the burner. Besides which, from its uneven-ness, the appearance of the flame was not so satisfactory. It was not until the principles which regulate the production of light from coal gas came to be known and observed in the construction of burners, that a modification of the old idea was arrived at, which enabled the benefits of a reduced pressure to be obtained without any of the attendant evils hitherto experienced.

The first real improvement of the union-jet burner.

A modification in the construction of the union-jet which, though slight, was nevertheless a real improvement, appears to have been made at an early period in the history of this burner. Instead of having the top of the burner perfectly flat, it was made slightly concave ; more especially at its centre, where the two jets of gas emerge. The effect of this alteration was to enable the stream of gas to spread out better ; and thus to cause the flame to become broader at its base. The shape of the flame was thereby improved ; and (what is of more consequence) its illuminating power increased, because air was not drawn so readily into the midst of the flame. The value of the arrangement is shown by the fact that it has been retained ever since, and is made use of in the latest and most improved burners of this class.

Prior to 1860, numerous novel contrivances were introduced as " improved " burners ; but all were not equally valuable with the simple arrangement just described. The construction of many of them, indeed, betrayed a lamentable ignorance of the first principles of gas combustion. For instance, one is described as " a fishtail with four converging holes ; and there is an aperture in the centre of the burner for the admission of atmospheric air into the flame ! " Another was a batswing with two or more slits, producing a series of flames amalgamated into one ; by which means it was supposed that an improved duty was obtained from the gas—unmindful, or, more probably, in ignorance of the fact that the same quantity of gas, properly consumed through one slit, would yield a better light.

The double-flame burner.

A burner which, at different times, and under various names, has been brought repeatedly into notice is the double-flame ; con-sisting of two batswing or union-jet burners set at an angle to each other, so that their flames converge, and merge into one. When two gas-flames are made to coalesce in this manner, a greater amount of light is developed than the sum of that yielded by the separate flames ; provided that, in the combined flame, the gas is properly consumed, without smoke. The reason for this increase is twofold. First, the increased quantity of gas burnt in one flame enables a higher average. temperature to be maintained; and, in addition, a smaller surface of flame is exposed to the cooling action of the atmosphere than when the same quantity of gas is consumed in two flames. Second, the pressure at which the gas burns is diminished, because the initial velocity with which the streams of gas issue from the two burners is expended in impinging against each other, and a thicker flame results; the apparatus being, as far as its effect is concerned, a union-jet burner on a large scale. The increase of light so obtained appears to have been noticed at an early period; as a burner embodying the same principle is described and figured in "Clegg's Treatise," published in 1848. In Clegg's burner the gas issued from two perforated parallel plates inclined to each other; but at a more recent period two fishtail burners were employed, being mounted on separate tubes which branched out to a short distance from each other. Occasionally, for experimental and show purposes, it has been constructed with the two branches hinged together, so as to show the different effects produced when the two burners are used separately and in combination. At the present day it is made, by various makers, as one burner with two nipples, as shown in the annexed illustration; which doubtless is its most perfect form.

FIG. 5.—DUPLEX BURNER.

The advantages of the double flame are not so obvious under the conditions which obtain at the present day as at the period when it was first introduced. The increase of light it affords is most apparent when the gas is being consumed at an excessive pressure. Although, in general, it may be taken that any two flames, when combined, will develop a higher duty, per cubic foot of gas consumed, than separately; yet it would appear that this is not so in every case. When the gas is being consumed at the critical pressure which gives the best results, the flames are so near the smoking point that the slight diminution of pressure experienced when the streams of gas impinge upon each other is sufficient to cause the combined flame to smoke. Moreover, to such a stage of perfection have the ordinary flat-flame burners now been brought, that, for all ordinary consumptions, it may be safely affirmed that equal, if not superior results can be obtained with a single as with a double flame. Where, however, larger quantities of gas are required to be dealt with than can be effectively consumed in a single burner, the principle of combining two or more burners together, so that their flames shall mutually assist each other, may be advantageously employed; as is seen in the combination of flat-flame burners in the large lamps now employed in improved street lighting.

Scholl's "Platinum Light Perfecter."

An ingenious device for improving the efficiency of union-jet burners was brought out some twenty years ago by a Mr. Scholl, of London, and known as Scholl's "Platinum Light Perfecter," which is shown in the accompanying illustration. It consisted of

FIG. 6.—SCHOLL'S PLATINUM LIGHT PERFECTER.

a little brass ring, carrying a plate of platinum about 0·4 inch long by 0·15 inch wide. The ring fitted on to the top of the burner in such a manner that the platinum plate was held, in a vertical position, between the two orifices from which the gas emerged. The jets of gas, instead of impinging upon each other, impinged against the plate, and united above to form the flame. By the interposition of the metal plate, the velocity of the gas was much reduced; and a thicker and more sluggish flame was produced, with the result of increasing its illuminating power. When the apparatus was used upon a burner having very small orifices, and delivering its gas at a high pressure, the increase of light obtained was very striking; but with lower pressures the advantage derived from its use was correspondingly diminished. This is very clearly shown by the following table, which is extracted from a report, by Captain

Webber and Mr. Rowden, on experiments upon gas-burners, carried out at the Paris Universal Exhibition, 1867.*

Kind of Burner.	Cubic Feet of Gas per Hour.	Pressure in Inches.	Illuminating Power.		Increase per Cent.
			Without Perfecter.	With Perfecter.	
Leoni's fishtail, No. 2 .	3	0·84	1·3	4·1	215
„ „ No. 3 .	3	0·46	2·4	4·6	91
	4	0·70	2·8	6·5	132
„ „ No. 4 .	3	0·31	3·4	5·0	47
	4	0·47	4·5	7·6	68
	5	0·71	5·0	9·2	84
„ „ No. 5 .	4	0·42	5·3	6·9	30
	5	0·60	6·1	8·3	36
	6	0·81	7·1	10·0	40[1]
„ „ No. 6 .	4	0·31	6·2	8·0	29[2]
	5	0·46	8·0	10·4	30[3]

[1] Flame flickers. [2] Do. [3] Flame flickers a great deal.

Burners were also made with the metal plate forming part of the burner head; and, instead of being of platinum, it was sometimes formed of thin steel, or other commoner metal. Where platinum was used, some advantage probably accrued from its becoming incandescent; but, of course, any benefit arising from this source was not obtained when steel was employed. The remarks which have been made respecting the limited applicability of the double-flame burner will apply, with equal force, to the apparatus under notice. Although it effected an undoubted improvement when applied to burners ill adapted to the pressure at which the gas was supplied, equally good results could be obtained without its aid, when a burner was employed suited to the quality and pressure of the gas supplied.

Perhaps the most efficient flat-flame burners available prior to 1867 were those made by Mr. S. Leoni, of London. One of these is shown in fig. 7. This maker produced both batswing and union-jets; various sizes being made of each burner. Besides affording fairly good results from the gas consumed, the burners were supplied at a very moderate price. Their distinguishing feature was the peculiar substance of which the burner tips were formed. This was a material invented by Mr. Leoni, and named by him " adamas." (The precise composition of " adamas " is a trade secret; but it appears to consist of a mixture of various minerals or earths, moulded in a clayey or plastic condition, and then burnt.) Previous to his invention, the tip of the burner, or the burner head,

Leoni's flat-flame burners.

had been made, almost exclusively, of iron or brass.   There were, however, some grave defects inherent in the use of metal for this purpose.   The orifices of union-jets and the slits of batswings in course of time became much obstructed by the corrosion of the metal; and the efforts made to remove the obstruction only served to destroy the burner more quickly, by increasing the size and injuring the precise shape of the apertures.   The "adamas" tips, on the other hand, perfectly withstood the high temperature to which they were exposed, were quite incorrodible, and were sufficiently hard to endure a considerable degree of even rough usage. By constructing the tip of this material, the efficiency of the burner was improved in many ways.   The liability of the burner to corrosion being removed, and the inconvenience due to this cause done away with, the life of the burner was prolonged, and the expense

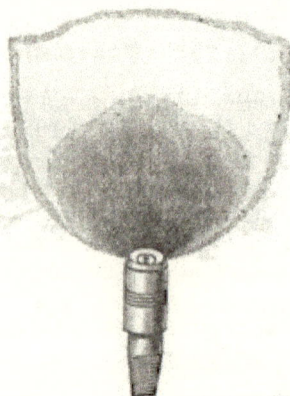

FIG. 7.—LEONI'S FLAT-FLAME BURNER.

of renewal consequently reduced.   But, in addition to these advantages, there was yet another direction in which the "adamas" tip contributed to enhance the utility of the burner.   This was in maintaining a higher temperature of the flame; and arose from its inferior capacity, compared with metal, for conducting heat from the flame.   That the advantage derived from this source, although unimportant, was not altogether imaginary, will be apparent when it is mentioned that metal burners, when in operation, usually attain to a temperature of from 400° to 500° Fahr.—an indication of the amount of heat being continuously abstracted from the flame. The adoption of a non-conducting material for the burner tip, while it did not entirely prevent, considerably reduced the loss of heat.

Two varieties of each class of burner were made by Mr. Leoni. In the one burner, the "adamas" tip was inserted into an iron stem; in the other, the tip was inserted in a brass body, which fitted on to the iron stem. Between the brass body and the iron stem of the latter burner there was affixed a layer of wool, designed to check the pressure at which the gas was supplied. Owing, very probably, to the unsuitability of the material (wool) used for this purpose, the result was not satisfactory; as, according to the statements of Messrs. Webber and Rowden, in the report previously cited, no difference could be detected, in many experiments, between the results yielded by the burner with or without the layer of wool. Some light is shed upon this apparent anomaly by certain experiments made by the writer to determine the pressure at which gas issues from various burners. With one of Leoni's No. 4 union-jets, under an initial pressure of 1 inch (the pressure at the inlet when the burner is in operation), the pressure at the outlet of the burner, when the layer of wool was employed, was 0·11 inch; but from the same burner, when the layer of wool was removed, the gas issued at a pressure of only 0·07 inch. Thus the effect of inserting the layer of wool in the burner was exactly the opposite of that which it was intended to produce; the pressure of the issuing gas stream being increased instead of diminished.

BRÖNNER'S BURNERS.

The credit of having produced the first flat-flame burners designed upon scientifically correct principles belongs undoubtedly to Herr Julius Brönner, of Frankfort-on-the-Maine. Long before the date of his invention, efforts had been made to reduce the pressure of the gas within the burner. But these endeavours were carried out in so hap-hazard a fashion as to lead to the belief that no definite conception was entertained as to what was really required. As we have seen, layers of wool had been employed; but the area of the interstices, or the gas-way through the material, was a matter of the merest accident. And there was not the slightest guarantee that the same conditions should prevail in any two burners. Herr Brönner shrewdly detected the cause of former failures, as he clearly perceived the end which it was requisite to attain, and towards which previous inventors had been but blindly groping. Having formed a right estimate of the requirements to be fulfilled, and the difficulties to be surmounted, he set about accomplishing the desired result by other means. There were two causes which had chiefly contributed to the unsuccessful issues of previous attempts. One was the uncertain and indefinite operation of the means employed for diminishing the pressure; the other was the inadequate provision for enabling the gas to lose the current, or swirl, acquired in passing the diminishing arrangement, and come to a state of comparative rest before issuing into the atmosphere. Both these errors were successfully avoided in Brönner's invention

—the former by making the inlet to the burner of restricted and
definite dimensions, and of less area than the outlet, or slit; the
latter by enlarging the chamber, or place of expansion within the
burner, as well as by the different arrangement adopted for diminishing the pressure.

Construction
of Brönner's
burners.
The general appearance of Brönner's burner is pear-shaped; and
in size it is considerably larger than an ordinary burner designed
to pass an equal quantity of gas. It consists of a cylindrical brass
body surmounted by a steatite top, and tapering to a very small
diameter at its lower end, or inlet; the latter being closed by a
plug of steatite, in which is a rectangular slot, or aperture, of
accurately defined dimensions. The size of this aperture deter-
mines the quantity of gas which, at any particular pressure, is
admitted to the burner; and the slit, or outlet of the burner, being
of greater area than the inlet, ensures the gas being delivered from

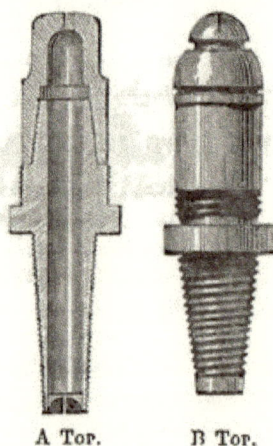

A TOP.   B TOP.
FIG. 8.—BRÖNNER'S BURNERS.

the burner at a lower pressure than that at which it enters it. By
varying the respective dimensions of these two openings, and their
relation to each other, the burner may be regulated to deliver its
gas at any required pressure short of the initial pressure at the
entrance to the burner. The enlargement of the cylindrical body
provides an expansion chamber, wherein the velocity of the stream
of gas which rushes through the narrow opening at the inlet of the
burner is checked, and any agitation or unsteadiness which may
have been imparted to it is subdued before the gas issues into the
atmosphere and is consumed. There are two kinds of tops for
the burners, which are distinguished by the letters A and B. The
B top is of the ordinary semi-spherical type, giving a true batswing-
shaped flame; the A top is flatter, almost square in form, and

yields a flame taller than, but not so broad as the former. In consequence of this difference in the shape of its flame, the latter burner is better adapted for use in globes. The general appearance of the burners, and their distinguishing peculiarities, will be clearly understood from the illustrations.

The material of which the more important parts of the burner are constructed is eminently adapted for the purpose. Steatite is a mineral which, as found in nature, is so soft as to be readily turned in a lathe, and shaped to any design; but when heated up to about 2000° Fahr. it becomes almost as hard and durable as flint, while perfectly retaining its form and colour. These properties peculiarly qualify it for receiving a slit or orifice, which, though of minute proportions, must be accurately formed to precise dimensions. Besides which, like " adamas," its capacity for conducting heat away from the flame is so limited that, in this respect, it has a considerable advantage over metal for the purpose of being formed into gas-burners. *Properties of steatite.*

The following tables, which are extracted from the report of the Committee of the British Association appointed to investigate the means for the development of light from coal gas of different qualities,* exhibit the very satisfactory results obtained by the use of these burners. In Table I., the gas operated upon was cannel gas (such as is generally supplied in Scotland), and possessed an illuminating power, when employed in the standard burner, of 26 candles per 5 cubic feet. Table II. contains the results of determinations with common gas (such as is used in London, and generally throughout the greater part of England); 5 cubic feet of which, in the standard burner, gave an illuminating power of 16 candles. The first and second columns of the tables refer to the different sizes of the tops and bottoms of the particular burners employed; there being in all some 16 sizes of the one, and 11 sizes of the other. These, being interchangeable, permit of a great variety of combinations; and enable a burner to be selected suited to any particular quality or pressure of gas. For as with pressure, so with illuminating power: In order to obtain the utmost lighting efficiency, different burners are required for gases differing in quality or their degree of richness. A burner which, with gas of one quality, will yield excellent results, may, under the same conditions of pressure and supply, be totally unsuited to gas of another quality. That this should be so will be evident from a consideration of what has been said as to the theory of burning gas to the best advantage; and, in brief, results from the richer gas containing in its composition a greater proportion of carbon, and so *Varied adaptability of the Brönner burner.*

---

* See *Journal of Gas Lighting*, Vol. XXXII., p. 423, and Vol. XXXVI., p. 376.

requiring an increased supply of air for its thorough combustion. This increased supply of air can only be obtained (with flat-flame burners) by causing the gas to issue into the atmosphere at a higher pressure; and so it comes about that, compared with the quantity of gas to be delivered through them, the slits of batswing and the orifices of union-jet burners must be considerably narrower when intended for cannel gas than when common gas is to be consumed. In other words, in order to develop its full illuminating power, it is essential that the pressure at which the gas issues from the burner should be proportioned to its quality. The gist of the matter is set forth in the general statement that "the poorer the quality of the gas, the lower must be the pressure at which it is consumed; and *vice versâ*."

TABLE I.

	At 1·0-Inch Pressure.					At 1·5-Inch Pressure.			
No. of Burner.	No. of Top.	Cubic Feet per Hour.	Illuminating Power.	Illuminating Power per Five Cub. Ft.	No. of Burner.	No. of Top.	Cubic Feet per Hour.	Illuminating Power.	Illuminating Power per Five Cub. Ft.
2	2	1·20	5·07	24·13	2	2	1·40	5·25	18·75
2	3	1·40	6·64	23·71	2	3	1·95	7·37	18·90
2	4	—	Smokes	—	2	4	2·30	10·33	22·46
2	5	—	,,	—	2	5	2·40	11·24	23·42
2	6	—	,,	—	2	6	—	Smokes	—
2½	2	1·40	5·53	19·75	2½	2	1·90	8·30	21·84
2½	3	1·70	8·48	24·94	2½	3	2·30	10·14	22·04
2½	4	2·03	10·33	25·49	2½	4	2·70	12·08	22·37
2½	5	—	Smokes	—	2½	5	2·85	14·29	25·07
2½	6	—	,,	—	2½	6	3·00	15·21	25·35
3	2	1·45	6·27	21·62	3	2	2·00	8·48	21·20
3	3	1·90	8·66	22·79	3	3	2·40	11·34	23·63
3	4	2·13	11·24	26·39	3	4	2·80	14·84	26·50
3	5	—	Smokes	—	3	5	3·15	17·04	27·20
3	6	—	,,	—	3	6	3·25	18·07	27·80
3½	2	1·50	5·81	19·36	3½	2	2·12	8·85	20·87
3½	3	1·95	8·30	21·28	3½	3	2·55	12·63	24·76
3½	4	2·55	12·08	23·68	3½	4	3·00	14·47	26·12
3½	5	2·80	14·38	25·68	3½	5	3·50	18·07	25·81
3½	6	3·00	15·58	25·97	3½	6	3·60	19·45	27·01
4	2	1·60	6·36	19·87	4	2	2·30	9·77	21·24
4	3	2·10	10·69	25·45	4	3	2·90	13·83	23·84
4	4	2·65	13·37	25·23	4	4	3·30	17·06	25·85
4	5	3·45	17·61	25·52	4	5	4·10	21·57	26·30
4	6	3·55	18·07	25·45	4	6	4·20	22·40	26·66
5	2	1·77	7·38	20·85	5	2	2·60	9·68	18·81
5	3	2·30	11·90	25·87	5	3	3·30	13·64	20·67
5	4	3·30	15·40	23·33	5	4	4·00	19·91	24·14
5	5	4·10	20·74	25·29	5	5	5·00	25·36	25·36
5	6	4·30	22·68	26·37	5	6	5·30	27·66	26·10

TABLE II.

No. of Top.	No. of Bottom.	At 0·5-Inch Pressure.			At 1·0-Inch Pressure.			At 1·5-Inch Pressure.		
		Cubic Feet per Hour.	Illuminating Power.	Illum. Power per Five Cub. Ft.	Cubic Feet per Hour.	Illuminating Power.	Illum. Power per Five Cub. Ft.	Cubic Feet per Hour.	Illuminating Power.	Illum. Power per Five Cub. Ft.
A 2	1	—	—	—	1·5	2·7	9·0	2·0	4·0	10·0
,,	2	1·6	2·9	9·1	2·4	5·2	10·8	3·1	6·8	11·0
,,	2½	2·0	3·9	9·8	2·9	6·8	11·7	3·8	9·4	12·4
A 3	3	2·1	4·4	10·5	3·2	7·8	12·2	4·4	10·6	12·0
,,	3½	2·5	4·8	9·6	3·8	9·2	12·1	4·9	12·2	12·4
,,	4	2·5	5·4	10·8	3·8	9·6	12·7	5·2	13·6	13·1
,,	4½	3·0	6·4	10·7	4·5	10·8	12·0	5·9	14·8	12·5
,,	5	3·2	7·7	12·0	5·1	13·2	13·0	6·8	18·0	13·2
,,	6	3·7	8·7	11·8	5·8	15·5	13·3	7·7	21·0	13·6
,,	7	3·5	8·6	12·3	5·9	16·0	13·6	8·4	23·0	13·7
,,	8	3·7	9·0	12·2	6·2	16·8	13·5	8·6	23·4	13·6
B 1	1	—	—	—	1·3	2·3	8·8	1·8	3·5	9·7
B 2	2	1·3	2·3	8·8	2·1	4·4	10·5	2·8	6·4	11·4
,,	2½	1·6	3·0	9·4	2·5	6·0	12·0	3·4	8·4	12·4
B 3	3	2·0	3·8	9·0	3·0	7·2	12·0	4·1	10·1	12·3
,,	3½	2·3	4·3	9·3	3·4	7·7	11·3	4·5	11·0	12·2
B 4	4	2·3	4·7	10·2	3·6	8·8	12·2	5·0	13·0	13·0
,,	4½	2·7	5·2	10·9	4·3	10·4	12·1	5·6	15·0	13·4
B 5	5	3·1	7·0	11·3	4·9	12·9	13·2	6·5	18·0	13·8
B 6	6	3·8	9·6	12·6	5·9	16·4	13·8	8·0	23·0	14·4
B 7	7	4·0	10·2	12·8	6·6	19·0	14·4	9·0	26·0	14·4
B 8	8	4·7	11·8	12·6	7·3	22·0	15·1	9·6	30·0	15·7

Doubtless the chief cause of the remarkable efficiency of the Brönner over previous burners is to be found in the pressure at which the gas flows from the burner and is consumed. In the course of some experiments made to determine the pressure at which gas is delivered from various burners, the writer found that from a No. 4 Brönner, with an initial pressure—*i.e.*, the pressure at the inlet when the burner is in operation—of 1 inch, the gas issued at a pressure of only 0·05 inch; and with an initial pressure of 0·5 inch, the outlet pressure was only 0·03 inch. On the other hand, a No. 4 steatite flat-flame burner, without any arrangement for retarding the flow of the gas, under the same initial pressure gave at the outlet 0·16 inch and 0·05 inch respectively. The absence of anything within the burner to cause the gas to swirl, or to issue with an unsteady flow, must also be credited with contributing, in no slight degree, to the favourable results yielded by these burners.

Pressure of gas with the Brönner burner.

THE HOLLOW-TOP BURNER.

In the hollow-top burner we have one of the most notable improvements which have been effected in flat-flame burners. A simple modification of the batswing—the earliest of flat-flame burners—it is not more complicated in its details than is that burner. Yet, simple as it is, its construction exhibits an important advance upon the original batswing. Indeed, this burner may be said to embody the only considerable improvement that has been made in the distinctive features of the batswing since the introduction of the latter burner, which, as we have seen, took place as early as the year 1816.

The hollow-top an improved batswing burner.

In its outward form, the hollow-top burner differs little, if at all, from the batswing; but a slight modification which has been adopted in the arrangement of its interior has produced a very marked result in improving the shape of the flame yielded by the burner, and, to some extent, in the results, as regards illuminating power, which it is capable of affording. In this burner, as its name implies, the interior of the top or head of the burner is hollowed out, forming an enlargement of the cavity or chamber within the burner, and (what is chiefly important) making the shell of the dome-shaped burner head of equal thickness throughout. In the ordinary batswing, in consequence of the varying thickness of the burner at this part, the slit is much deeper in the middle than at any other part of its length, and gradually decreases in depth towards each end. As the resistance to the passage of the gas, or the friction which it encounters, increases with the depth of the slit, the gas passes out from the burner at the ends of the slit more readily than in the middle; producing a wide-stretching flame, of scanty height in proportion to its width. From the same cause the flame is not of equal thickness throughout; being thinner in the middle than at the ends. Moreover, the outer extremities of the flame, extending so far beyond the body of the burner, are unable to retain the form given to them by the lateral flow of the gas at the ends of the slit; the resistance, presented by the atmosphere, together with the natural tendency of the gas to ascend, causing the under portion of the flame to fold back upon itself. As one result of this combination of untoward circumstances, the flame is liable to smoke with a slight agitation of the surrounding air.

In the hollow-top burner, the slit is of equal depth throughout its length; and the resistance offered to the passage of the gas being the same in all parts of the slit, the gas flows through the middle as readily as at the ends—nay, in reality rather more so, owing to the innate ascensive power of the gas, due to its being lighter than air. The peculiar hollowing-out of the head of the burner, also, withdraws the ends of the slit out of the direct course or current of the gas through the burner; so that the tendency of the stream of gas to issue at these points, in preference to going through the middle of the slit, is further checked. The consequence is that the shape of the flame is considerably improved; it being taller, more compact, and not so broad as that of the batswing. In addition, the flame being of equal thickness throughout, its illuminating power is somewhat improved; while, from its compactness, it is better enabled to resist atmospheric influences. With this alteration in the shape of the flame all original resemblance to a batswing is entirely destroyed; but the appearance of the flame of the new burner is much more agreeable to the eye than that of the older batswing.

As has been exemplified in so many instances in the history of invention, the hollow-top burner was not appreciated at its true value until long after it had been brought into existence. It appears to have been originally invented by Joseph and James Wadsworth, of Marple and Salford, and was patented by them in

Who invented the hollow-top burner.

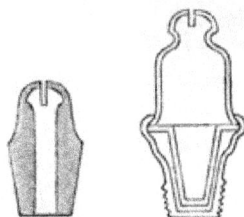

FIG. 9.—ORIGINAL HOLLOW-TOP BURNER.

(From Wadsworth's Specification.)

1860. According to the specification of the inventors, the burners might be made either in solid or sheet metal, as will be seen from the accompanying illustrations, copied from the drawings in the specification. But there were difficulties in the way of casting the burners in solid metal which do not seem to have been surmounted; and those produced under the patent appear to have been made exclusively of sheet brass. For many years these burners were made and sold without their peculiarities attracting any marked attention; which would seem to imply that their faulty construction precluded the attainment of all the advantages afforded by the burner as we know it.

The superior results which the hollow-top burner was calculated to afford did not become fully apparent until the burner was made of non-conducting material, and greater care exercised in its construction. This appears to have been done in Germany earlier than in this country. But, in England, it was undoubtedly Mr. Sugg who first turned his attention to the improvement of the burner, and demonstrated its merits. Mr. Sugg commenced the

manufacture of this burner in steatite in the year 1868; and since
that time the burner has been extensively employed, and its
advantages widely recognized. The superiority of hollow-top
burners produced by Mr. Sugg to those previously manufactured,
is chiefly the result of their being made in steatite instead of
in metal. With this material, greater exactness and uniformity
are obtained in the shape and dimensions of the burner than
when metal is employed; besides which there is (what has been
before referred to) the advantage arising from its inferior con-
ductive capacity for heat, and its non-liability to corrosion.
Another improvement, also due to Mr. Sugg, and which is productive of noticeable results, consists in cutting the slit of the
burner with a circular saw, applied from above, so as to make the
ends of the slit curved instead of horizontal; by which means the
tendency of the gas to issue laterally at the ends of the slit, and
form horns to the flame, is lessened. Mr. Sugg's table-top burner
(which was introduced in 1880), in addition to the characteristic

1868 BURNER. 1874 BURNER. TABLE-TOP BURNER.
FIG. 10.—SUGG'S HOLLOW-TOP BURNERS.

features of the hollow-top, has a. rim-like projection from the
burner, below the slit; its object being to protect the flame from
the disturbing influence of the uprush of air in its immediate
vicinity, and so preserve its shape unaltered, while diminishing its
liability to smoke. Prior to Mr. Sugg—namely, in the early part
of 1879—Mr. Bray had successfully obviated this injurious action
upon the flame of the ascending current of air, by affixing to the
burner two arms of brass, so placed as to be immediately under
the projecting wings of the flame.

BRAY'S BURNERS.

The burners of Messrs. George Bray and Co. have deservedly
acquired a world-wide reputation, and are in extensive use wherever
gas lighting is known. Their distinguishing characteristic, and
that which has won for them the high repute in which they are
held, is the union of cheapness with remarkable efficiency. In all

the various descriptions and classes of burners which are produced
by this firm, the characteristic referred to is preserved; although it
is needless to add that the different varieties are not equally efficient. Of the three forms of flat-flame burners we have been considering—batswing, union-jet, and hollow-top—the one which, more
than any other, has been the speciality of the firm is the union-
jet; and it is with the development of this class of burner that the
name of Bray is most intimately and honourably associated.

The "regulator" union-jet, which was the first notable burner
produced by Messrs. Bray, has received, perhaps, a wider application than any other single gas-burner. It consists of a cylindrical
brass case, screwed at one end for insertion into the fittings, and at
the other containing a tip of "enamel"—a material invented by
Mr. Bray, and apparently of somewhat similar composition to the
"adamas" of Mr. Leoni—the "enamel" tip being perforated, in
the usual manner, with two holes, set at an angle to each other, for
the outflow of the gas. The distinctive feature of this burner is the
introduction into the lower part of the brass case of a layer, or

Bray's
"regulator"
burner.

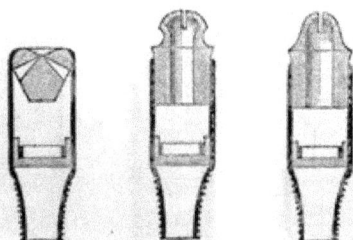

UNION-JET. HOLLOW-TOP OR SLIT-UNION.* BATSWING.

FIG. 11.—BRAY'S "REGULATOR" BURNERS.

layers, of muslin; designed to check in some degree, and to steady
the current or flow of the gas through the burner. At the time of
its introduction, this burner compared very favourably, as regards
the results it yielded, with other burners in common use; and its
fairly good performances, together with the very low price at which
it can be sold, cause it still to be extensively employed wherever
the attainment, from the gas consumed, of the highest obtainable
results may be subordinated to cheapness, or in situations where,
from delicacy of construction or from the care and attention demanded, a more efficient burner may not be so suitable. But

Bray's
"special"
burner.

* The name "slit-union," by which Mr. Bray prefers to designate this
burner, he states to be derived from the resemblance of its flame to that
of the union-jet burner; while it is produced by means of a slit.

in the matter of developing the illuminating power of the gas employed, the "regulator" is far surpassed by the more recently introduced " special " burner of the same makers.

Mr. Bray's series of "special" burners—embracing union-jet, hollow-top, and batswing—are constructed upon the principle of, and in form are somewhat similar to Brönner's burners, which have already been fully described. Apart from its being of greater bulk, the main divergence in the construction of the "special" burner from that of the earlier "regulator" is the introduction, into the lower part of the brass case, of a plug or washer of enamel, pierced by a small circular hole for the admission of gas into the burner; the diameter of this hole determining the quantity of gas which, at any particular pressure, is admitted into the burner. Just above the enamel washer, a layer of muslin is inserted, as in the "regulator" burner; which, in this case, is for the purpose of subduing the agitation, or swirl, acquired by the current of gas in passing through the narrow aperture in the washer. A tip of enamel, made of the particular description (union-jet, hollow-top,

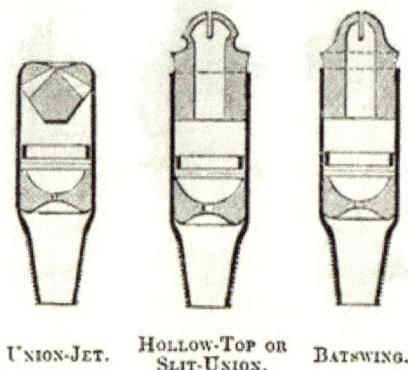

UNION-JET.   HOLLOW-TOP OR SLIT-UNION.   BATSWING.

FIG. 12.—BRAY'S " SPECIAL " BURNERS.

or batswing) required, fitting into the upper part of the brass case, completes the burner. The objects aimed at in the "special" burner are to cause the gas to be consumed at the lowest pressure compatible with the maintenance of a firm flame, and with the least agitation, or swirl, in the current of gas as it issues from the burner. The former is attained, as in Brönner's burners, by diminishing the area of the opening admitting into the burner, without a corresponding diminution of the orifices through which the gas issues into the atmosphere; the latter, by the interposition of the layer of muslin which is immediately above the diminishing arrangement, as well as by the enlargement of the gas chamber in the upper part of the burner. The improvement thus effected in the illuminating power developed from the gas is well shown in the following tables extracted from an exhaustive series of tests of gas-burners carried out by Mr. T. Fairley, F.R.S.E., Borough Analyst of Leeds, and embodied by him in a report presented to the Leeds Corporation. The full text of the report will be found in the *Journal of Gas Lighting* for February 6, 1883.

Medium Lighting Power Union-Jets.

	"Regulator" Burners.					"Special" Burners.			
No. of Burner	Pressure in Inches	Cubic Feet per Hour.	Illuminating Power in Stand. Candls.	Illuminating Power per 5 Cubic Feet.	No. of Burner	Pressure in Inches.	Cubic Feet per Hour.	Illuminating Power in Stand. Candls.	Illuminating Power per 5 Cubic Feet.
3	0·5	3·50	6·8	9·7	3	0·5	3·43	11·3	16·4
3	1·0	4·80	6·9	7·2	3	1·0	4·90	15·6	15·8
3	1·5	6·20	7·5	6·05	3	1·5	6·03	17·6	14·6
4	0·5	4·65	12·2	13·1	4	0·5	3·73	13·3	17·8
4	1·0	6·67	14·2	10·6	4	1·0	5·15	17·4	16·9
4	1·5	8·16	14·2	8·8	4	1·5	6·57	22·4	17·1
5	0·5	5·72	17·0	14·9	5	0·5	4·80	17·6	18·3
5	1·0	7·97	20·0	12·6	5	1·0	6·67	24·4	18·3
5	1·5	9·73	21·8	11·2	5	1·5	8·30	30·0	18·2
6	0·5	5·90	18·0	15·2	6	0·5	5·48	20·1	18·3
6	1·0	8·35	23·0	13·8	6	1·0	7·65	28·4	18·6
6	1·5	10·60	28·0	13·2	6	1·5	9·20	34·2	18·7

Medium Lighting Power Slit-Unions.

	"Regulator" Burners.					"Special" Burners.			
No. of Burner	Pressure in Inches	Cubic Feet per Hour.	Illuminating Power in Stand. Candls.	Illuminating Power per 5 Cubic Feet.	No. of Burner	Pressure in Inches.	Cubic Feet per Hour.	Illuminating Power in Stand. Candls.	Illuminating Power per 5 Cubic Feet.
3	0·5	4·22	13·8	16·4	3	0·5	3·04	10·8	17·8
3	1·0	6·37	20·2	15·9	3	1·0	4·61	16·4	17·6
3	1·5	8·14	25·8	15·9	3	1·5	5·88	19·9	16·9
4	0·5	4·25	14·8	17·4	4	0·5	3·82	14·2	18·6
4	1·0	5·88	20·6	17·5	4	1·0	5·69	20·8	18·3
4	1·5	7·95	26·5	16·6	4	1·5	7·35	25·6	17·5
5	0·5	5·25	19·0	18·2	5	0·5	4·12	15·4	18·7
5	1·0	8·14	28·4	17·45	5	1·0	6·37	23·4	18·4
5	1·5	10·20	36·4	17·8	5	1·5	7·94	28·5	18·0
6	0·5	5·67	22·2	19·6	6	0·5	5·00	19·6	19·6
6	1·0	8·60	33·6	19·4	6	1·0	7·55	29·0	19·2
6	1·5	11·10	39·5	17·8	6	1·5	9·70	37·0	19·1

D

Medium Lighting Power Batswings.

	"Regulator" Burners.					"Special" Burners.			
No. of Burner	Pressure in Inches.	Cubic Feet per Hour.	Illuminating Power in Stand. Candls.	Illuminating Power per 5 Cubic Feet.	No. of Burner	Pressure in Inches.	Cubic Feet per Hour.	Illuminating Power in Stand. Candls.	Illuminating Power per 5 Cubic Feet.
3	0·5	4·16	12·6	15·1	3	0·5	3·37	12·4	18·4
3	1·0	5·64	16·6	14·8	3	1·0	5·25	20·4	19·4
3	1·5	7·83	21·0	13·4	3	1·5	7·13	24·0	16·8
4	0·5	4·26	14·0	16·4	4	0·5	3·67	13·0	17·7
4	1·0	6·74	21·2	15·6	4	1·0	5·55	20·6	18·6
4	1·5	7·81	24·0	15·3	4	1·5	7·13	26·0	18·2
5	0·5	4·76	15·4	16·2	5	0·5	3·86	14·6	18·9
5	1·0	6·93	20·4	14·7	5	1·0	5·85	22·6	19·4
5	1·5	8·72	25·8	14·7	5	1·5	7·53	28·0	18·6
6	0·5	6·04	20·0	16·5	6	0·5	4·86	19·4	20·0
6	1·0	8·82	29·4	16·6	6	1·0	7·53	31·6	21·0
6	1·5	11·10	31·6	14·2	6	1·5	9·60	39·0	20·4

The quality of the gas operated upon averaged about 19 candles when tested with the Standard London Argand Burner.

In a former part of this treatise it was remarked that the flames produced by the modern representatives* of the batswing and fish-tail burners have lost the original resemblance to the objects whence the names of those burners were derived; and that the two flames have gradually approached each other in shape, until, in their latest developments, they are practically identical. We have seen how that, by the invention of the hollow-top, a burner is obtained apparently, to all outward appearance, the same as the batswing, yet giving a greatly improved form of flame. We have now to learn how the fishtail, or union-jet burner has been modified so as to yield a flame closely agreeing with that produced by the improved slit burner.

How the union-jet burner has been improved.

As first constructed, the union-jet burner gave a tall, narrow flame; its extremity being forked and jagged like the tail of a fish. Besides being unsightly, this form of flame was ill-adapted to develop, to anything like its full extent, the illuminating power of the gas. In order to obtain the best results, as regards illuminating power, the heat-intensity of the flame must be very high, so as to bring up the temperature of the particles of carbon liberated in the flame to the necessary degree of incandescence. To this end there must be concentration of the flame, in order to utilize to the full the heat of combustion. With the tall flame produced by the

---

* Although the true batswing is still in common use, I look upon the hollow-top as being its "modern representative;" seeing that, in a great many instances, it has superseded the former burner—of which, indeed, it is only an improved form.

original union-jet burner there was too much exposure to the atmosphere for the flame to attain to the requisite intensity of heat; as
well as considerable liability of the gas being brought too early
into intimate contact with air, and so oxidized, or fully consumed,
before its carbon had been raised to the temperature necessary
to enable it to give out light. With the burner in its improved
form the height of the flame is much curtailed, while it is broadened,
and made more even and compact. This alteration has been
chiefly brought about by two modifications in the construction of
the burner-tip—first, by hollowing out its flat upper surface; and,
second, by altering the angle at which the two streams of gas
emerge from the burner. By scooping out the central portion of
the flat top of the burner, so as to form a hollow or depression
where the gas emerges, the flat sheet of flame which is formed
when the two streams of gas impinge upon each other obtains a
broader base, and at the same time is preserved from drawing air
into its midst. But the chief share of the improvement is due
to the alteration in the angle formed by the two channels in the
burner-tip. It will be readily apparent that the more obtuse this
angle—that is, the nearer the two streams of gas are to impinging
against each other in a horizontal line—the more will the flame
tend to spread out, or the lower the pressure required to obtain any
desired spread of flame. It is by taking advantage of this circumstance that Mr. Bray has been enabled to improve the union-jet
burner. Twenty years ago this burner was usually made with the
two channels in the burner-tip placed at an angle of about 60°. In
Bray's "regulator" burner, introduced in 1869, they were placed
at an angle of 90°; with the result of obtaining a more satisfactory
flame, both as regards its appearance and illuminating power. In
the "special" burner, which was not brought out till 1876, the
angle is increased to 120°; thus enabling the necessary spread of
flame to be obtained with the gas issuing at a low pressure.
Another minor improvement in the latter burner consists in making
the holes in the burner-tip elliptical instead of circular.

CHAPTER III.

ARGAND BURNERS.

The premier position among gas-burners undoubtedly belongs to the Argand ; and it is from no unwillingness to recognize its claims, much less from ignorance of its merits, that I have left the consideration of this burner until now. It occupies this honourable position as much by virtue of the importance it has acquired through being accepted by Parliament as the test burner, and the peculiar relation in which it consequently stands to other burners, as for any marked superiority in operation. For while, in general, the Argand gives superior results to other burners, this is not always the case. There are circumstances and conditions to which the Argand is quite inapplicable, and where a simpler and less pretentious burner will give excellent results. Indeed, some of the simple flat-flame burners which we have had under notice have now been brought to such a stage of perfection, that, when intelligently used, they not unsuccessfully rival the Argand. But it has been in the direction of demonstrating the illuminating power which it was possible to obtain from gas, and stimulating to the attainment, by other and simpler burners, of the same level of excellence, that the influence of the Argand has been most beneficial. For, by reason of its peculiar construction, and more especially its mode of obtaining the air necessary for combustion, the Argand lends itself, more readily than any other burner, to the work of investigating and experimenting upon the conditions necessary for economical combustion, and the development of the highest illuminating power from the gas consumed. In this burner, the air supply to the flame is under complete control ; and thus one of the chief elements of uncertainty and difficulty which are experienced in dealing with other burners is eliminated. The delivery of gas to different parts of the flame is also more susceptible of variation ; and the results of such variation more fully exposed to observation. The consequence has been that the most remarkable advances in developing improved illuminating power from coal gas have been made with this burner. But after the possibility of obtaining an improved duty from the gas has been demonstrated by means of the Argand, and the conditions necessary for its attainment determined, equally good results have been achieved by other burners.

In thus showing the benefits to be derived from a more scientific mode of combustion, and leading the way to the fuller attainment, by other burners, of the illuminating power obtainable from the gas, the Argand burner has acted as a pioneer in the development of gas lighting. For, on account of its complexity, and its delicacy of construction, this burner has never been, nor, indeed, can ever hope to be generally employed. Besides the inconvenience and expense entailed by the cleaning and renewal, when broken, of the

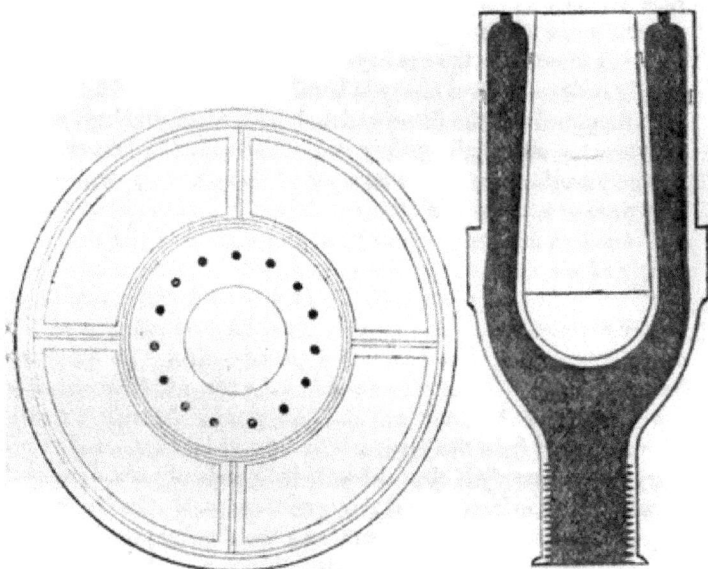

PLAN OF GLASS-HOLDER AND BURNER TOP. SECTION OF BURNER.

FIG. 13.—ARGAND BURNER.

glass chimney which is indispensable to this burner, its very perfection as a burner precludes its being adopted under the conditions which appertain to the great majority of situations in which gaslight is required. For while, under the particular conditions as to pressure of gas, &c., for which it has been constructed, the Argand may give results surpassing any other burner, a very slight divergence from these conditions is productive of far more damaging results to the illuminating power of the flame than is the case with other and less efficient burners. The cause of this seeming anomaly will be apparent when we come to consider in detail the construction of the Argand, and the conditions which must be observed to ensure its satisfactory operation. For the present it will suffice merely to make mention of what appear to be well-established facts—viz., that the most perfect burners are the least adapted for use under uncertain and varying conditions ; and that in proportion to the efficiency of a burner, under the conditions for which it has been constructed, is the injury to the illuminating power of its flame which is experienced when these conditions are departed from.

What is an
Argand
burner ? Resolved into its simplest form, the Argand burner may be said to consist of a hollow ring of metal, or other suitable material, provided with the necessary tubes or connections for communicating between its interior and the gas supply, and perforated on its upper surface with a number of holes for the emission of the gas. Through these holes the gas issues in a series of jets, which immediately coalesce to form one cylindrical sheet of flame. The burner is surmounted, and the flame enclosed, by a glass chimney, which is supported on a light gallery connected with the burner ; the chimney serving the double purpose of shielding the flame from draughts, or currents of air (thus enabling the gas to burn uniformly and steadily), and of drawing upon the surface of the flame the supply of air necessary for its proper and complete combustion. For in the Argand the air supply is produced under conditions totally different from those which govern its production in all the other burners we have had under consideration. In flat-flame burners, the quantity of air supplied to the flame is determined by the pressure of the gas ; or, in other words, the velocity with which it issues from the burner. In Argand burners, on the contrary, the air supply is obtained quite independently of the pressure at which the gas issues ; and the conditions most effective for the economical combustion of the gas, and the development from it of the highest illuminating power attainable, are only secured when the pressure of gas is reduced to a minimum.

It has been shown, in speaking of flat-flame burners, how the illuminating power of the flames yielded by such burners is injuriously affected by an excess of pressure in the gas, as it issues into the atmosphere, causing a too great intermingling of gas and air. With such burners, however, some degree of pressure is needed, in order, by bringing the flame into contact with sufficient of the oxygen of the air, to promote the requisite intensity of combustion ; whereas with the Argand the draught that is produced through the agency of the glass chimney enables the necessary supply of air to be obtained for the support of the flame without adventitious aid from the pressure of the gas. Consequently, one of the chief objects to be aimed at in the construction of the latter burner is to so reduce the pressure of the gas within the burner that it may issue with little or no greater velocity than that due to its own specific lightness. In some of the best Argands this object is attained very successfully; and the ingenious devices which have been made use of to gain this end will be duly described in the sequel. But, in addition to causing the gas to issue from the burner at the minimum of pressure, it must be delivered evenly and equally at all parts of the ring of holes; so that there shall not be an excess of gas supplied to one portion of the flame, and an insufficiency to others. Then the area of the opening in the centre of the ring, through which the air supply is obtained to the inner surface of the flame, as well as the length and diameter of the glass chimney, must be so proportioned that the exact quantity of air needed to enable the flame to yield its maximum results shall be drawn upon it. These, and other equally essential requirements, have to be taken into consideration, and provided for, in constructing an efficient Argand burner. It is no wonder, therefore, that the development of the powers of this burner has taken up so much time and labour and inventive skill; and the remarkable degree of efficiency to which it has now been brought testifies to the thought and the accurate knowledge of the principles of combustion which have been brought to bear upon it.

It is, however, only within comparatively recent years that its true principles of construction have been at all fully recognized, as evinced by the burners which have been produced. For a long period, Argand burners were made upon wholly empirical and arbitrary rules. During the early years of gas lighting, the makers of gas apparatus, and such persons as professed to have a special knowledge of the production and utilization of the new illuminant, appear to have been ignorant of even the most obvious of the conditions required for the successful working of the burner. In one of the earliest works which appeared relating to gas lighting,[*] we find the Argand burner described as consisting of "two concentric tubes closed at the top with a ring having small perforations, out of which the gas can issue; thus forming small distinct streams of light." According to this description, the burner referred to cannot have been an Argand in the strictest sense of the word; but, in reality, must have consisted chiefly of a series of single jets placed in a circle, and surrounded by a glass chimney. But the great improvement in the amount of light developed, which resulted from bringing the jets of flame closer together, so as to cause them to coalesce and produce one homogeneous mass of flame, could not long escape notice; and accordingly we find that in "Clegg's Treatise," which appeared twenty-five years later, the proper disposition of the holes in the ring, necessary for the successful

The earliest Argands

---

* Accum's "Treatise on Gas-Lights."

operation of the burner, is clearly recognized. In this work,
speaking of the Argand burner, it is remarked (p. 193) that "the
distance between the holes in the drilled ring should be so much
that the jet of gas issuing from each shall, when ignited, just
unite with its neighbour."

Before a really efficient burner could be produced, there were,
however, to be successfully encountered other problems, the precise nature of which was not so clearly apparent as that of the one
above referred to; otherwise their solution would not have been
so long delayed. Of these, the most important, and at the same
time the most difficult, were two—namely, the right adjustment
of the air supply, and the most advantageous pressure at which to
consume the gas. In the earliest Argands, not the slightest provision was made for diminishing the pressure of the gas before it
was consumed. It was thought that everything had been accom-
plished that was necessary if the holes for its emission were sufficiently minute to allow of no more than the required quantity of
gas passing through them at the extreme pressure at which
it was supplied to the burner. The consequence was that the
gas, issuing from the burner at a very high velocity, became so
intermingled with air before it was consumed, that its flame was
excessively cooled; and only a small fraction of the illuminating
power available was developed. Then as to the air supply. In
nearly every burner produced prior to Mr. W. Sugg's invention of
the "London" Argand in 1868, this was greatly in excess of the
requirements; nor is it to be wondered at. Had the supply of
air been delicately adjusted, while yet there was no provision for
diminishing the pressure of gas at the burner, the flame would
have been liable to smoke on any sudden increase in the pressure
of gas in the mains; and the annoyance and inconvenience occa-
sioned by a smoking flame were greater drawbacks than the loss of
light experienced through having the air supply greatly in excess.
Thus, although during this period there were many so-called
"improved" burners brought into notice, in none of them were
these two cardinal requirements in the production of an efficient
burner clearly recognized and seriously grappled with; and, consequently, the high level of excellence to which the Argand is capable
of being brought was not attained.

SUGG'S ARGANDS.

The 'London'
Argand.

The invention by Mr. W. Sugg, in 1868, of the famous "London"
Argand constitutes an important epoch in the history of gas lighting. Prior to that time, the construction of this class of burners
had been carried out in a wholly empirical manner; and such

improvements as had been effected must be looked upon as being rather the fortuitous issues of hap-hazard endeavours, than as resulting from the acquirement of clearer views as to the conditions to be complied with in order to ensure the successful operation of the burners. The invention of the " London " Argand was the first earnest attempt to abandon the former chance methods, and to proceed upon more scientific lines. Its construction shows that its inventor possessed a thorough acquaintance with the principles of combustion; while, in many particulars, it exhibits an intelligent discernment, and a successful application of the precise means required to attain a desired end. In this burner, the extreme importance of causing the gas to issue at a low pressure is for the first time clearly recognized; and the manner in which this object is so successfully attained is as simple as it is ingenious. At the entrance to the burner the gas is divided among three narrow tubes, the combined capacity of which is much smaller than that of the pipe supplying the burner. Through these tubes the gas is conducted into a concentric cylindrical chamber (forming the main body of the burner), where its rapid flow is checked; the current, or swirl, which it may have acquired, is subdued; and the gas comes to a state of comparative rest before it issues into the atmosphere and is consumed. The top rim of this concentric cylinder is pierced with 24 holes, the aggregate area of which is considerably greater than that of the three supply-tubes; thus ensuring that the gas shall be delivered at a much lower pressure than that at which it enters the burner. By dividing the gas into three streams, which enter the cylindrical chamber at equidistant points in its circumference, the supply is equally distributed throughout the entire ring of holes; and a flame of even and regular shape is the result.

The arrangement by which, in this burner, the air supply is obtained and regulated is as noteworthy as are the means adopted for controlling the pressure of the gas. The opening within the circular ring of holes is much smaller than in previous Argands; thereby proportionately reducing the quantity of air supplied to the inner surface of the flame. The space between the cylindrical body of the burner and the glass chimney is occupied by a truncated cone of thin metal, the upper edge of which is on a level with, and reaches to within a very short distance of the rim of the burner; while its base rests upon the gallery supporting the chimney. By means of this cone, all the air entering between the burner and the chimney is directed upon the immediate surface of the flame; thereby promoting intensity of combustion, and a higher illuminating power of the flame. Then the chimney itself is of such dimensions that, with the quantity of gas for which the burner has been constructed,

just sufficient air is drawn upon the flame to completely consume
the gas by the time the top of the chimney is reached; a flame of
such length as to nearly reach to the top of the chimney, without
smoking, being the most effective and economical for the quantity
of gas consumed.

Another matter which tended not a little to enhance the results
yielded by this burner was an alteration in the material of which
the body of the burner was constructed.   In previous Argands,
this had, in almost every instance, been metal; whereas in the
" London " burner steatite was employed.   How the illuminating

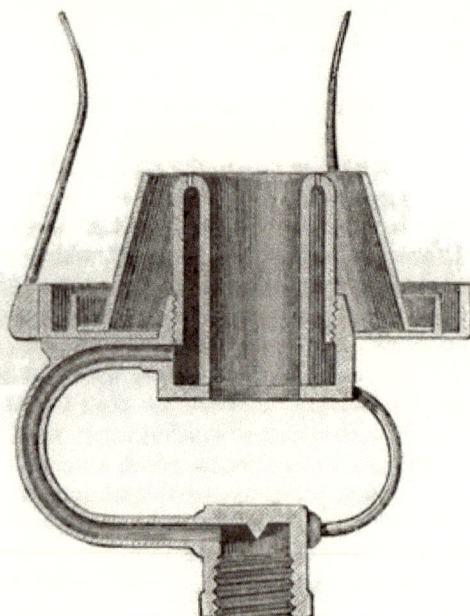

FIG. 14.—SUGG's " LONDON " ARGAND.
(*Full Size.*)

power of the flame is affected by the material of which the burner
is constructed has been gone into so fully before (in relation to
flat-flame burners), that it is unnecessary to dwell upon the matter
here; only remarking that as in Argands the contact surface
between the burner and the flame is relatively so much greater
than in flat-flame burners, the cooling of the flame due to this
cause is proportionately increased.

So great was the improvement effected by this burner in the
illuminating power developed from the gas consumed, so obvious
its superiority to every previous Argand, that it was immediately

adopted by the Metropolitan Gas Referees as the standard burner The standard test burner. for testing ordinary coal gas within the area of their jurisdiction; and from that time down to the present it has continued to be prescribed in Acts of Parliament as the burner to be employed in testing ordinary coal gas, not only in the Metropolis, but generally throughout the United Kingdom. But although, as the standard test-burner, the original "London" Argand can still be obtained, it has been far surpassed, in the results yielded, by a new series of Argands, in which the same ingenious inventor has still further applied the principles first put into practice in the former burner. In this newer series of burners, the details of construction before The improved "London" Argand. adopted are modified in two or three particulars; but without departing from the general principles embodied in the arrangement of the earlier burner. Thus the holes in the ring are considerably larger, while the three supply-tubes remain of exactly the same capacity as before; by which means the gas is delivered at a much lower pressure. As the increased size of holes necessitates that the cylindrical body of the burner should be of enlarged diameter, the opening in the centre becomes of greater area than before. Were it to remain so, it would permit too large a quantity of air to be drawn upon the inner surface of the flame; to obviate which result a metal spike rises in the centre, reducing the area of the opening, and proportionately diminishing the quantity of air which would otherwise be admitted at this part of the burner. The arrangement for regulating the air supply to the outer surface of the flame is likewise modified, but in a different direction. The upper edge of the cone is brought nearer to the rim of the burner, and slightly curved, so as to direct the air more completely upon the flame; while the base of the cone, instead of extending to the glass chimney in an unbroken surface, is pierced by a number of holes, which admit air between the cone and the chimney. The action of this third current of air is to keep the chimney cool, and to steady the flame; and, in addition, it may be that it provides a supply of air to support and intensify combustion at the upper extremity of the flame. The combined effect of these alterations is to cause the burner to develop from 7 to 12 per cent. more light from the gas consumed, than is yielded by the original "London" Argand.

The Silber Argand, which is a remarkably efficient burner, in Silber's Argand burner. the main features of its construction is very closely related to Mr. Sugg's later Argands just described. The air is directed on to the outer surface of the flame, as in those burners, by a curved deflector, of which the upper edge is, however, at a higher level than in Mr. Sugg's burners. Air is also admitted between the deflector and the glass chimney. The most striking divergence in its construction from that of Mr. Sugg's burners is contained within the opening in the centre of the burner. Instead of a solid metal spike, there is a brass tube, through which, as well as between its circumference and the cylindrical body of the burner, air can enter to feed the inner surface of the flame. In addition to promoting the steadiness of the flame, it would appear that the air entering through this inner tube supports the combustion of the gas at the tail of the flame. The arrangements for diminishing the pressure of the gas within the burner, and for ensuring its equable distribution to all parts of the ring of holes, though quite different, seem to be scarcely less complete than those employed in the "London" burner. From the nipple which connects the burner to the gas supply, the gas enters (by four minute perforations) into a horizontal chamber, where its velocity is checked, and whence it is conveyed into the cylindrical chamber forming the main body of the burner. The very satisfactory performances of the burner (which are in advance of those of the standard Argand) sufficiently attest the correctness of its construction.

Multiple Argands.

For consuming large quantities of gas, double or treble Argands are constructed. These consist, in effect, of two or three Argand burners placed concentrically to each other within one chimney. Mr. Sugg has produced a series of burners of this class, designed to pass quantities of gas ranging from 15 to 55 cubic feet per hour; and, in some instances, exceeding even the latter figure. These burners, with ordinary (16-candle) coal gas, give a light equal to 4 candles per cubic foot of gas consumed; which is a considerably better result than is afforded by the standard burner. The cause of their yielding results so superior to the ordinary Argand is found in the circumstance that their flames present a much smaller surface area to the cooling action of the air, in proportion to the quantity of gas consumed. The arrangement of these burners differs from that of the improved single Argands, which have been described, only in that there are two or more steatite cylinders, each fed by its own supply-tubes, and having its own distinct ring of holes; while the space between the cylinders is so proportioned as to admit no more than the quantity of air required to produce the necessary intensity of combustion.

THE DOUGLASS BURNER.

The multiple or concentric Argand invented by Mr. (now Sir) J. N. Douglass, the Engineer to the Trinity House, may be mentioned here. This burner is of the type of those last noticed, but possesses certain peculiar features which give it a distinct claim to novelty. As will be seen by the accompanying illustration, the concentric cylinders of which the burner is composed terminate at different heights; their tops forming a regular gradation of steps, of which the innermost is the highest. These cylinders are of considerable depth, permitting the gas and air to be heated by contact with their surfaces before the point of ignition is reached. The essential feature of the invention, however, is a series of deflectors of peculiar shape, which, in addition to directing air on to the surfaces of the flames, are so formed "as to force the outer flame or flames on to the inner flame or flames in the manner illustrated." By this means the flames are concentrated and united into one, and combustion is

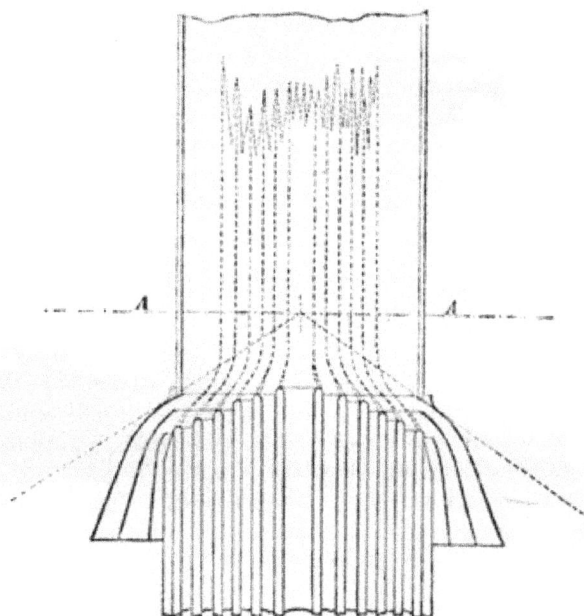

Fig. 15.—The Douglass Argand.

(*A A, Focal Plane, or Belt of Strongest Light.*)

quickened; and, a greater intensity of heat being thus attained, the illuminating power is much augmented. When this burner was first brought into notice, in 1881, high hopes were entertained as to its future. The results which it was said to afford, being far in advance of anything previously obtained from a simple Argand, seemed to promise for the burner a speedy and unequivocal success. At the North-East Coast Marine Exhibition, held in 1882, a burner with ten rings was exhibited, which was reported to develop, from 16-candle gas, 6 candles per cubic foot—a truly remarkable result to be given by so simple a burner. But, notwithstanding its apparently successful introduction, the burner has made little or no headway in the direction of its practical application. Indeed, it may almost be said to have faded altogether out of public view. This would seem to imply that there are difficulties in the way of its successful working, when brought under ordinary conditions, which were not foreseen at the time of its invention.

CHAPTER IV.

GOVERNOR-BURNERS.

Throughout this treatise, much has been said of the relation which the pressure of gas, at the point of its delivery from the burner, bears to the illuminating power of the flame yielded—sufficient to show that the maintenance of a low and equable pressure in the gas supply is one of the conditions most imperative to be observed for the attainment of economy in combustion. Ordinarily, however, this condition does not obtain at the consumers' burners. The exigencies of distribution require that, in order to maintain a sufficient supply wherever gas is needed, a much higher pressure should be kept in the mains than is requisite for developing, at the burner, the best results from the gas consumed. Moreover, the pressure at any one point is subject to continual fluctuations from the variations in the consumption of gas going on in the neighbourhood. For instance, where a number of burners are in operation in a house, consuming about the exact quantities of gas for which they have been constructed, when part of them are shut off the gas supply to the remainder is in excess of what is required; and, consequently, the burners do not develop the same proportion of light from the gas consumed as formerly. Where a large consumption of gas is suddenly discontinued (as in the business parts of a town, when the shops and warehouses are closed), the increase of pressure that is experienced at the burners which remain in operation is very manifest. The effect of this increase in the pressure of the gas supply is seen in different directions in Argand and flat-flame burners. In the former, it causes the flame to smoke, by permitting more gas to pass through the burner than can be properly consumed; in the latter, by cooling the flame below the temperature required for effective combustion, it reduces, in proportion to the extent to which it is higher than the original pressure, the illuminating power developed per cubic foot of gas consumed.

Effects of excessive pressure with Argand and flat-flame burners.

Seeing that economy in combustion can only be attained under the conditions of an equable pressure, it becomes necessary to subdue the fluctuations above referred to, or at least to prevent their reaching the burner. To this end the regulator, or governor, is employed. In this instrument, a bell dipping into, and sealed

The gas regulator.

in liquid—or else a flexible leather diaphragm—is actuated by the pressure of the entering gas, and so connected with a valve as to reduce the area of the opening which permits gas to enter the instrument in proportion to the pressure of gas at the inlet; by which means an equable pressure is maintained at the outlet, no matter what the quantity of gas which is being consumed, or how the pressure may vary in the inlet-pipe. By the aid of a governor, fixed on the service-pipe at the entrance to a building, the pressure of gas at the various burners is rendered fairly uniform; yet, even then, perfect equality of pressure is not obtained. The slight friction which the gas experiences in flowing through the pipes causes the burners to be supplied at somewhat lower pressures, the farther they are removed from the burner. And, again, owing to its low specific gravity, gas tends to gain in pressure with an increased elevation; each rise of 10 feet adding about 1-10th of an inch to its pressure. From this cause a higher pressure is experienced in the upper than in the lower rooms of a building. This peculiarity was observed at an early period in the history of gas lighting; as Clegg mentions that, in cotton-mills, check-taps were employed to regulate the pressure of gas at each floor.* In order, therefore, to obtain the desired regularity of pressure in the gas supply, governors must be employed for each storey; or, what is better still, each burner must have its own separate governor. And this brings us back to the subject with which we are more closely concerned.

The governor-burner, as its name implies, consists of a governor, as described above (but, of course, on a smaller scale) combined with a gas-burner; the governor being adjusted so as, whatever excess of pressure there may be in the gas-supply pipes, to permit only the quantity of gas to pass which the burner is intended to consume. Obviously, the principle herein contained is capable of receiving numerous applications. It can be, and is applied with equal success to Argand and flat-flame burners; while the modifications which obtain in the manner of constructing the regulating portion of the apparatus are almost as numerous and as varied as are the burners themselves. As the main features exhibited by one are common to all, it is unnecessary to go into the details of their several constructions. It will suffice to take two or three of the most successful, or the best known, as representatives of the whole.

Giroud's Rheometer.

Among the first in order of time—and still retaining no unworthy position in order of merit—is the " rheometer," or " flow-measurer," of M. Giroud. In this instrument a light metal bell is sealed in glycerine contained in a cylindrical case; the bottom of this latter containing the inlet-pipe, screwed for connecting to the ordinary fittings, while from the centre of its cover rises a tube leading to the burner. The bell is pierced by a small hole for the passage of the gas, and is surmounted by a cone-shaped projection, which constitutes the valve of the instrument. As the pressure of the entering gas lifts the bell, it causes this cone-valve to enter the mouth of the tube leading to the burner; reducing the area of the opening in proportion to the pressure of gas acting upon the under side of the bell, and so permitting only the required quantity of gas to pass to the burner. It might be thought that the presence of liquid would constitute an objection to the use of the instrument;

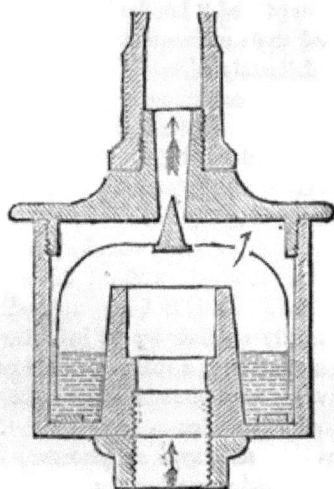

FIG. 16.—GIROUD'S RHEOMETER.

but, as glycerine does not evaporate, when once the instrument is fixed and properly adjusted, it needs no further attention. With an excessive initial pressure, there is, however, a liability of the gas to bubble through the sealing liquid, and so destroy the efficiency of the instrument; but this might be obviated by increasing the depth of the bell, and so giving it a greater seal. The instrument is very reliable for the purpose which it is intended to fulfil; delivering, through a considerable range of pressure beyond that required to raise the bell, the exact quantity of gas for which it has been adjusted. It may be added that the rheometer has an advantage over many instruments of its class, in that it presents so little obstruction to the downward rays of the flame.

Mr. William Sugg, in his regulator or governor, adopts an entirely different arrangement to the foregoing. The valve is

E

placed at the inlet of the governor; and not at its outlet, as in the instrument just described. Instead of a metal bell, a diaphragm of thin and very flexible leather is employed, which is raised by the pressure of the entering gas, and, in turn, actuates the valve; closing the entrance to the governor in proportion to the pressure of gas acting upon it. The orifice communicating between the under and the upper side of the leather diaphragm is controlled by a screw, whereby the quantity of gas delivered to the burner can be regulated according to requirements; but when once it has been adjusted to give any desired pressure of gas at the burner, this pressure will be strictly maintained, no matter with what excess of pressure (within reasonable limits) the gas may be supplied to the instrument. The improved "London" Argands produced by Mr. Sugg (the details of the construction of which have been already described) are too delicately adjusted to be applied with advantage directly to the ordinary consumer's gas-fittings, or wherever any variation in the pressure of the gas supply is likely to be experienced. However, with the addition to them of the above governor, their use becomes as easy and simple as that of other burners; and thus the gas consumer is enabled to obtain the benefit of the most improved apparatus without being called upon to exercise the constant care and attention which, without the aid of the governor, would be necessitated. Besides being applied to Argands, this governor is successfully applied by its inventor to his flat-flame burners. In conjunction with a simple steatite burner of the latter class, it has received a very extended application, under the name of the Christiania governor-burner.

Recently, however, a new type of governor, for application to burners, has been brought out by the same manufacturer, the construction of which is very different to that of the instrument referred to above; and as it is somewhat simpler in its details, and withal appears to be cheaper in construction, it seems destined to supersede the former instrument. In this new governor, instead of a leather diaphragm, there is a bell (or float) of steatite, which is free to move, in the manner of a piston, within an inner cylindrical chamber contained within the outer case of the instrument. Attached to the centre of the float, and on its upper surface, is a tube sliding within another tube of somewhat larger area; the latter forming a continuation of the inner cylindrical chamber. The smaller tube is open at both ends, and thus communicates from below to above the float; the outer tube is closed at the top, but has an orifice in its side. The action of the instrument is as follows: —The gas, entering below the float, passes through the inner tube to the upper part of the cylindrical chamber, and thence, through the orifice in the outer tube, to the burner. As the pressure of the

entering gas exceeds that required to overcome the weight of the
float, the latter is raised; the tube which is attached to it being
propelled farther into the outer tube in which it slides, and, in so
doing, partially closes the orifice in the side of the latter. In this
way, according to the pressure of the gas acting upon the under
side of the float, the area of the opening through which it must
flow to get to the burner is reduced; and so the quantity of gas
which issues from the burner remains the same under all pressures
above that required to actuate the float. The instrument appears
to be as reliable as it is simple, and to contain few parts calculated

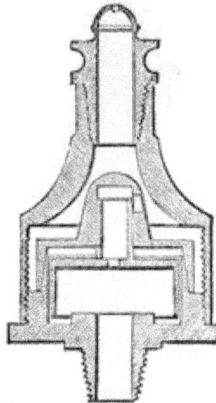

FIG. 17.—SUGG'S STEATITE-FLOAT GOVERNOR.

to get out of order; but, of course, whether or not it will retain
its good qualities after long-continued use can only be proved by
experience.

Another instrument of this class—the last which I shall notice—
is Peebles's needle governor-burner. For simplicity combined with
remarkable efficiency, it is undoubtedly ahead of all its compeers.
Somewhat similar in principle to Giroud's rheometer, it differs
from that instrument in many of the details of its construction;
and while dispensing with the use of liquid, maintains equal efficiency
in operation. It was described as follows by Dr. W. Wallace, in
a lecture on "Gas Illumination," delivered before the Society of
Arts in January, 1879 :*—"In a little cylinder stands a so-called
needle, on the point of which rests a flanged cone of exceedingly
thin metal. At one side of the cylinder there is a small tube leading away the gas, and the orifice of which is influenced in area by
the action of the cone. The instrument, by means of a screw
leading into the side tube, can be made to deliver any desired

*Peebles's
needle
governor-
burner.*

* See *Journal of Gas Lighting*, Vol. XXXIII., p. 162.

number of cubic feet, which it does with surprising accuracy, provided that the pressure of the gas is not less than 6-10ths of an

Efficiency of the needle governor-burner.

inch." As to the efficiency of the instrument, Dr. Wallace proceeded to state :—" In trials that I have made, I have not found the variations of volume at different pressures to exceed 1 per cent." For situations where this extreme nicety of operation is not absolutely essential, or where the rate of consumption is to be

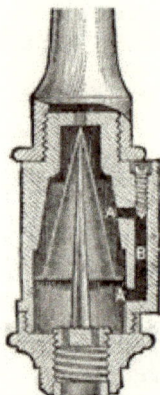

FIG. 18.—PEEBLES'S NEEDLE GOVERNOR.

invariable, the instrument is constructed in a somewhat modified and simpler form. The small tube on the side of the instrument is dispensed with, and the gas permitted to pass through perforations in the lower part of the cone. With this alteration there is a nearer approach to the construction of the rheometer; but, as in that instrument, there is no provision for altering the rate of consumption to suit different circumstances.

CHAPTER V.

REGENERATIVE BURNERS.

As was remarked in the introduction to this treatise, recent years have witnessed a very considerable advance in the construction of gas-burners, and in the amount of light capable of being developed from each cubic foot of gas consumed. Undoubtedly the most noticeable feature of this advance is the successful application of the regenerative, or, as it would be more appropriately designated, recuperative system. Briefly stated, this consists in utilizing the heat of the products of combustion from the gas-flame (which otherwise would be dissipated into the atmosphere) to raise the temperature of the gas before it is ignited ; and, likewise, of the air necessary for combustion. The temperature of an illuminating gas-flame is usually estimated to be between 2000° and 2400° Fahr. ; and as the products of combustion must leave the flame at a temperature little, if at all, inferior to the former figure, it must be evident that there is an ample margin of heat for employment in this direction. A considerable proportion of the large amount of heat conveyed by those products of combustion which, under ordinary circumstances, is imparted to the surrounding atmosphere—often elevating its temperature to an unnecessary and prejudicial extent—is, by this method, returned to the flame ; intensifying the process of combustion, and augmenting, in a remarkable degree, the illuminating power developed from the gas consumed. Thus the ultimate effect of the operation is to produce a concentration of heat in the flame, and the conversion of superfluous heat into beneficial light. Within a comparatively recent period, the utility of this process was strongly disputed ; and it was stoutly maintained, by many persons, that as the immediate effect of ignition was to cause a temperature of more than 2000° Fahr. to be attained, the heating of the gas and air prior to their combustion could produce little or no beneficial effect upon the illuminating power of the flame. However, the falsity of this view of the case is conclusively demonstrated by practical experiment ; the remarkably high results yielded by burners that have been constructed upon the regenerative system sufficiently attesting the correctness of the principles upon which they are founded.

Temperature of a gas-flame.

Although, in general, both the gas and air supplies are heated, it is chiefly due to the latter that the beneficial effect noticed is produced; and this for two reasons. First, because the quantity of air is so much greater than the gas it is required to consume; being, at the nearest approach to theoretical perfection, fully six times its volume. Second, because four-fifths in volume of the air consists of inert nitrogen, which does not contribute anything to the heat of the flame, but, when applied in its normal, cold condition, abstracts no inconsiderable proportion of heat from it. Yet the heating of the gas itself is not without very appreciable influence. In an ordinary gas-flame there is always an area of non-illumination around, and extending to a variable distance from the burner-head. This is caused partly by the conduction of heat from the flame by the burner; but, in a greater degree, by the cooling action of the issuing stream of cold gas, as is shown by its extending farther from the burner in proportion to the pressure or velocity with which the gas issues. The prejudicial effect due to the former is obviated to a great extent by constructing the burner of steatite, or other non-conducting material. To remedy the latter, nothing will avail but the heating of the gas supply.

Effects of heating the gas and air. The effect of heating the gas is to enlarge the area of the illuminating portion of the flame, and, in a minor degree, to enhance the intensity of incandescence to which the carbonaceous particles are raised. When the gas issues from the burner at a temperature little inferior to the temperature of ignition, the hydrocarbons it contains are immediately decomposed; the liberated particles of carbon are raised to the temperature of incandescence; and the illuminating area of the flame is extended downwards, even to the surface of the burner. The heating of the air operates chiefly to produce and maintain a more elevated temperature of the flame; and, in this manner, contributes to the development of a higher illuminating power from the same area of flame. In the case of ordinary gas-flames, the cold atmosphere by which they are surrounded, by abstracting heat from the flame, prevents the most favourable conditions for the development of light from being attained. When, however, the air immediately surrounding the flame has been previously heated, the particles of carbon (the incandescence of which furnishes the desired illuminating power) attain to a much more exalted temperature; and, consequently, give out a greater degree of light.

But there is yet another direction in which the prior heating of the air supply contributes to the development of improved illuminating power. Being heated, its density is lowered; so that in any given volume of air there is less weight of oxygen than when cold. The consequence is that as less oxygen is presented to a given surface-area of flame, the separated particles of carbon remain for a longer period of time in the incandescent condition before being entirely consumed. Thus there are three distinct results produced by heating the gas and air before combustion—namely, first, the particles of carbon are liberated earlier in the flame; second, they are raised to a more exalted temperature; and, third, they remain for a longer time in the incandescent condition. The combined effect of all three is the improved illuminating power developed from the gas consumed.

So far back as the year 1854, the principle of heating the air supply to an Argand burner, by means of waste heat from the flame, was partially applied, with some success, by the

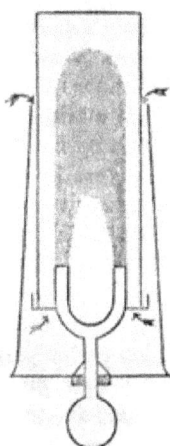

FIG. 10.—BOWDITCH'S REGENERATIVE
GAS-BURNER.

Rev. W. R. Bowditch, M.A., of Wakefield. Mr. Bowditch's burner, which is shown in the accompanying diagram, contained, in addition to the ordinary chimney, an outer glass chimney, which extended for some distance below the inner one, and was closed at the bottom; so that all the air needed to support the combustion of the gas was required to pass down the annular space between the chimneys, and in its passage became intensely heated by contact with the hot surface of the inner chimney, as well as by radiation from the flame itself. This burner contained many defects. Amongst others, the inner chimney could not long withstand the intense heat to which it was subjected, and, in consequence, had to be frequently renewed; the heating of the air was not effected solely by the products of combustion, but, perhaps in a greater degree, by the abstraction of heat from the flame itself;

Bowditch's regenerative burner.

while, at best, this heating was but partial. Yet, these defects notwithstanding, the burner showed very clearly the beneficial results attending even a partial application of the principle; as, in the illuminating power it developed from the gas consumed, a clear gain of 67 per cent. over the ordinary Argand burner was obtained. Although the drawbacks connected with the construction of Mr. Bowditch's burner prevented its ever receiving general, or even extensive adoption, its simplicity has gained for it the distinction of being freely copied by so-called inventors of a later day.

It was left to Herr Friedrich Siemens, of Dresden, to produce a burner which, while applying the principle of regenerative heating in the most scientific and complete manner, should also be adapted to the ordinary conditions of gas lighting. After much experimenting on the subject, a burner embodying the essential features of the regenerative system was invented by this gentleman in 1879; and so great was the advance which its performances manifested over anything previously attained, so wide the prospect of further achievements which was opened out, that it may fairly be said to have inaugurated a new era in gas illumination. In this burner the products of combustion were made to give up a considerable portion of their heat to the gas and air, as the latter passed to the point of ignition; the flame itself not being called upon to contribute in any degree to this result. Although, as was but natural, the first attempts towards the construction of such a burner were very crude, and but partially successful in their results, the inventor persevered in his endeavours to work out his ideas into practical and thoroughly satisfactory shape. It was not until after it had gone through many modifications that the burner acquired the peculiar form which now distinguishes it, and attained to its present stage of perfection. Before proceeding to describe an example of the burner as now constructed, it is necessary to state that the principles embodied in Herr Siemens's invention are equally well adapted—and, indeed, are applied with equal success —to the construction of flat-flame and Argand burners; but as the distinctive features of the invention are common to both classes of burners, it will be quite sufficient to describe in detail one of the latter type.

A prominent feature in the appearance of the Siemens burner, as will be seen from the annexed illustration, is a large metal chimney, for creating a draught to carry away the products of combustion. The entrance to this chimney is situated a little above the apex of the flame; but there is a branch flue connecting the main chimney with the interior of the burner. The body of the burner is of metal, and its interior is divided into three concentric chambers. Of these, the innermost is open at the top, and is sur-

(marginal note:) Invention of the Siemens regenerative burner.

mounted by a porcelain cylinder, which, when the gas is lighted, is surrounded by the flame. This chamber is closed at the bottom, but communicates at the side with the before-mentioned branch.

ELEVATION.   ENLARGED SECTION OF COMBUSTION CHAMBERS.

FIG. 20.—SIEMENS'S REGENERATIVE GAS-BURNER.

tube, or flue, leading to the main chimney. The intermediate chamber communicates, at its lower extremity, with the gas supply; and terminates, a short distance from the top of the burner, in a number of small metal tubes, which convey the gas to the point of ignition. The outer chamber is open both at top and bottom, and is for conveying air to support the combustion of the gas. In order to promote greater intensity of combustion, there is a notched deflector at the summit of the latter chamber, and another on the lower part of the porcelain cylinder, which cause the air to impinge more directly upon both sides of the flame. There is also an arrangement for introducing air between the outer casing of the air chamber and the glass chimney which encloses the flame; its object being to keep the chimney cool.

The action of the burner is as follows:—When the gas is ignited at the ring of tubes, the heated air and products of combustion, which rise from the flame, create a draught in the main chimney. Through the communication established by means of the lateral flue, a partial vacuum, or area of low pressure, is induced in the innermost chamber of the burner, and within the porcelain cylinder

Action of the Siemens burner.

which surmounts it. As the flame terminates close to the mouth of the latter, the greater portion of the products of combustion, instead of going into the main chimney, are sucked into the porcelain cylinder; and thus a current is set up through the interior of the burner, and by the lateral flue, to the main chimney. The heat carried away by the products of combustion is communicated, through the walls of the chambers, to the entering gas and air; and by this means the latter are heated to a very high temperature before they issue from the burner and are consumed. The consequence is that a much greater intensity of combustion is maintained; the carbon particles are separated earlier in the flame, and are raised to a more exalted temperature; and the ultimate effect is a higher yield in illuminating power per cubic foot of gas consumed. Independent tests by various experienced photometrists have conclusively shown that a light equivalent to that from 5 to 6 candles is obtained per cubic foot, from gas which, in the standard "London" Argand, yields a light of only from 3 to $3\frac{1}{2}$ candles.

Defects of the Siemens burner.

While the advantages of the Siemens burner are many and obvious, it is not without its disadvantages. These partly arise from causes connected with the very observance of the conditions necessary to secure the efficiency of the burner. With every advance in the more efficient operation of gas-burners, increased care and attention are demanded in their employment, in order to obtain the benefits they are calculated to yield. Indeed, it would almost appear that the nearer the approach to perfection which is made in the construction of a burner, the greater must be the drawbacks to its general adoption. Thus, in the burner under notice, if the gas supply is allowed to become in excess, the tail of the flame enters the porcelain cylinder, and soot is deposited in the interior of the burner; obstructing the passages, and impairing the burner's action. Then, to cause the burner to yield its highest results, it is necessary that the air supply be accurately adjusted to the quantity of gas being consumed. To this end the entrance to the air chamber, at the bottom of the burner, is covered by a perforated semi-circular cup, by turning which the quantity of air entering the burner can be increased or diminished as required. Moreover, the bulky construction of the burner, with its accompaniment of chimney and flue, and its complicated arrangement of tubes and chambers, imparts to it a somewhat clumsy and inelegant appearance, which is calculated to impair the favour with which its remarkable performances cause it to be regarded. But these drawbacks are far outweighed by the undoubted advantages conferred by the burner —in improved illumination combined with economy of combustion, and the facilities it affords for securing perfect ventilation.

Encouraged by the success of Herr Siemens, other inventors have

followed in his footsteps; with the result that there are now a
variety of burners before the public, embodying the same principles,
but differing in the details of their construction and in the measure
of their efficiency. Of these may be mentioned Grimston's, Thorp's,
and Clark's; and without describing in detail the construction of
the several burners (of which further particulars will be found in
the "Register of Patents" in the *Journal of Gas Lighting**), it
must suffice to refer to the salient points and distinctive features
of each.

Grimston's burner (shown on the next page) consists, in effect, of Grimston's
an Argand burner turned upside down; the gas issuing from the regenerative
bottom ends of a number of small tubes placed in a circle. The burner.
jets of flame—first directed downwards from the mouths of these
tubes—by a conoidal deflector in the centre of the ring, are caused
to spread outwards, and assume a horizontal direction; and by
their amalgamation with each other a continuous sheet or ring of
flame is produced. The horizontal direction of the flame is maintained by its passing underneath a metal flange, faced with white
porcelain, or other refractory material; the supply of gas being
adjusted so that the flame just terminates at the outer edge of this
flange. Before entering the chimney, the products of combustion
are caused to flow through a number of vertical tubes contained
in a cylinder, which is concentric to an inner cylinder containing
the gas-supply tubes. The outer cylinder is traversed by the air
needed for the support of combustion, which is to become heated
before reaching the point of ignition; and in order the more completely to enable the products of combustion to impart their heat
to the entering air, the cylinder is further intersected by strips of
wire gauze, which pass around and between the tubes (see fig. 22,
on next page). By these means the air is intensely heated; and,
passing among the narrow burner tubes through which the gas is
conveyed, gives up a portion of its heat to the latter before the
point of ignition is reached. Thus, in a very simple manner, both
air and gas are raised to a considerable temperature before combustion takes place.

With regard to the efficiency of the burner, at the exhibition of
gas appliances held at Stockport in 1882 (where a gold medal was
awarded to it, as well as to Thorp's burner, to be referred to here-
after), with a consumption per hour of 9·84 cubic feet of 17·5
candle gas, an illuminating power of 60·67 candles was obtained
(equal to 6·16 candles per cubic foot); while, on another occasion,
when the burner was consuming 8·94 cubic feet per hour, an illuminating power of 51·5 candles (equal to 5·76 candles per cubic foot)

* See Vol. XL., pp. 786, 950; and Vol. XLII., p. 836.

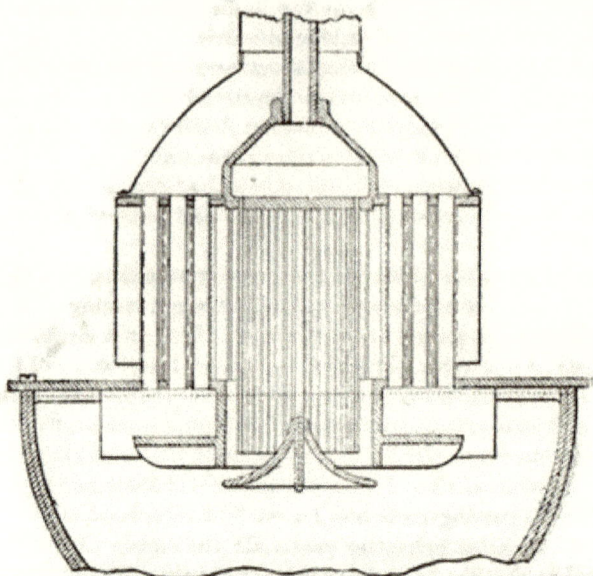

FIG. 21.—GRIMSTON'S REGENERATIVE GAS-BURNER.

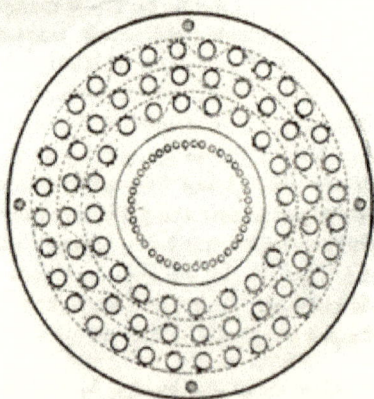

FIG. 22.—GRIMSTON'S BURNER.
PLAN, SHOWING REGENERATING ARRANGEMENT.

was obtained from gas of the same quality. It is claimed for this
burner that equally good results are obtained with small sizes as
with large; and this, if borne out in actual practice, should
go far towards ensuring the success and extensive adoption of the
burner.

Thorp's burner produces a cylindrical flame, like that of the Argand, but without the aid of a glass chimney which is a necessary adjunct to the latter burner. By means of a deflector on the inner side of the flame, the latter is made to curve outwards and assume a somewhat convex form, so as to obviate the shadow which otherwise would be cast by the gas chamber at the bottom of the burner. Above the flame is a cylindrical chimney, divided by a vertical partition into two concentric chambers, which are intersected by a series of metal gills, or projections, continued through both chambers. The outer chamber is for conveying away the products of combustion; the inner one for the passage of air to feed the flame; while down the centre of the inner chamber there passes a tube conveying the gas to the point of ignition. The hot products of combustion pass up from the flame

Thorp's regenera burner.

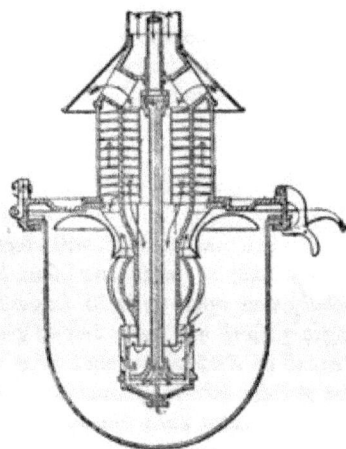

Fig. 23.—Thorp's Regenerative Gas-Burner.

through the outer chamber, and give up the greater portion of their heat to the projections; by which it is conducted into the inner chamber, and transferred to the incoming air. A common imperfection of regenerative burners is that, in consequence of the diminished rate at which the gas flows through the burner when expanded by heat, when starting the burner the gas must be only partially turned on, and the quantity gradually increased as the burner becomes heated; thus necessitating considerable attention. To prevent the need for this attention, there is in Thorp's burner an ingenious contrivance for automatically regulating the quantity of gas admitted to the flame. The central gas-tube, which is referred to above, contains a brass rod, fixed at one end, and at the other connected to a valve controlling the quantity of gas that enters the tube. At first, when the gas is lighted, this valve is almost closed; but as the rod becomes heated it elongates, gradually opening the valve until the full quantity of gas is admitted which the burner is intended to consume. At the Stockport exhibition, Thorp's burner was tested with the following results, as recorded in the Judges' report. After it had burned about two hours, "it gave an illuminating power of 183 standard candles, while burning 27 cubic feet of gas per hour (equal to 6·77 standard candles per cubic foot), with gas of 3·5 candles per cubic foot. . . . In another experiment with the same quality of gas, after burning half an hour it yielded, under similar conditions, 154 candles with a consumption of 25·29 cubic feet per hour, which gave an illuminating power of 6·02 candles per cubic foot."

Clark's regenerative burner. There is nothing in Clark's burner that calls for special notice. In its main features it appears to be constructed upon similar lines to Grimston's burner, although the coincidence is doubtless only accidental.* It must, however, be added that in the details of its construction it is much simpler than the latter burner; and certainly it appears to lose very little in efficiency from its greater simplicity, as the following extract from a report by Mr. F. W. Hartley, the well-known photometrist, will show:—" With a consumption rate of 5·3 cubic feet of gas per hour, the amount of light yielded horizontally was equal to 29·79 times that of a standard candle. The light yielded per cubic foot of gas burned per hour was therefore equal to 5·62 times that of a standard candle." And the amount of light delivered immediately downwards is said to be " very sensibly greater than the amount of light delivered horizontally." Like the Grimston burner, it is of the inverted Argand form; the gas issuing from a chamber at the bottom of a tube which descends through the centre of the burner. The products of combustion escape through a chimney; and in so doing give up a portion of their heat to the entering air, which is conveyed to the point of ignition through horizontal tubes that intersect the chimney. The burner is enclosed in a suitable lantern, the lower half of which consists of a semi-globular glass; a similar arrangement being adopted in connection with the Grimston and Thorp burners.

---

* In justice to Mr. Clark it should be mentioned that, since the above appeared in the *Journal of Gas Lighting*, the attention of the writer has been called to the fact (which had been overlooked by him) that Clark's patent was taken out some months before that of either Grimston or Thorp.

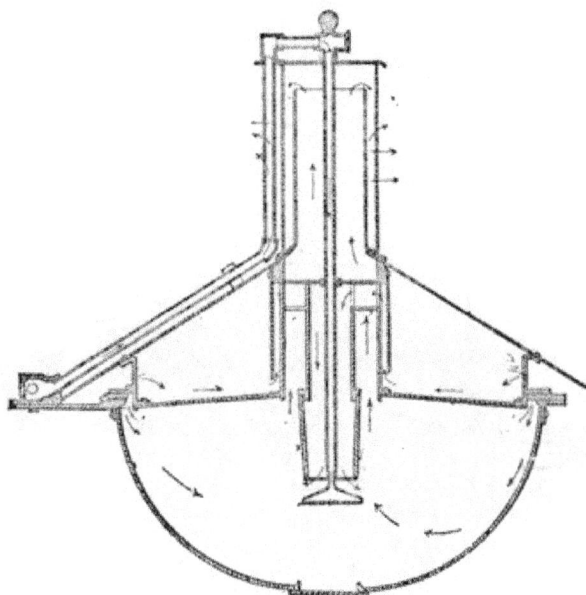

FIG. 24.—CLARK'S REGENERATIVE GAS-BURNER.

The three burners last mentioned have not been before the public sufficiently long to enable a reliable opinion to be formed as to their value in actual and prolonged use. Although there is no reason for supposing that such will occur in the present instance, it so often happens that the results indicated by apparatus in the experimental stage, or while still under the control of the inventor, are not borne out in practice, that it would be unwise to express any decided opinion as to their ultimate worth from existing information. It is, however, to be earnestly hoped that the marked favour with which they have been received will not be impaired on improved acquaintance; but that further experience will justify the anticipations that have been excited by the excellent performances of the burners hitherto, and demonstrate at once their durability and real usefulness.

Since writing the above, considerable activity has been shown by inventors in producing new burners upon the regenerative principle, or in improving upon existing models. Of course, as yet it is too early to arrive at a satisfactory estimate of their actual value or relative worth; but it may be hoped that, from the increased attention being devoted to the subject, some real and practical results will flow, by which the gas-consuming public will be the gainers.

So far, the most promising of this class of burners that has been brought into actual use, since the introduction of the Siemens burner, is the one represented below.

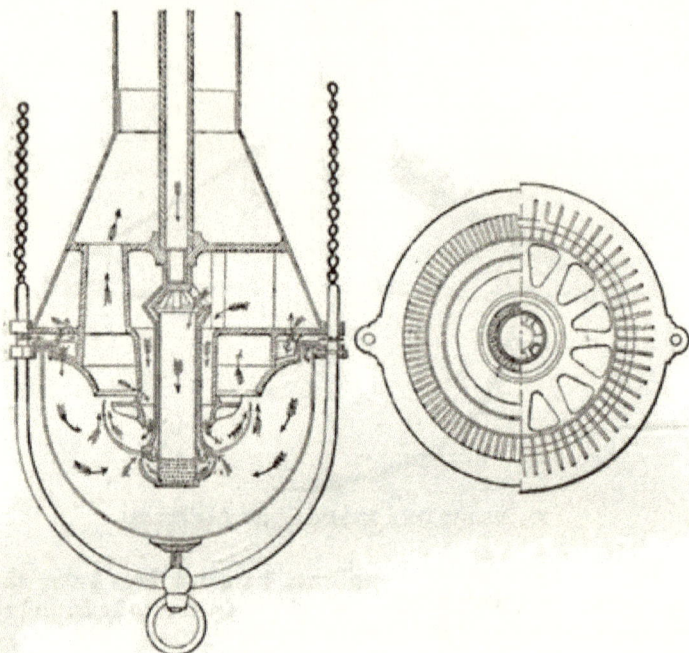

Fig. 25.—Bower and Thorp's Regenerative Gas-Burner.

It is a modification, in the direction of greater simplicity, of Thorp's former burner, illustrated and described on p. 69 of this treatise; and as its construction is based upon the same lines as that burner, further description is not required.

CHAPTER VI.

INCANDESCENT BURNERS.

A review of gas-burners would scarcely be complete without some reference to the incandescent burners of M. Clamond and Mr. Lewis. Although their dependence upon an artificially produced blast or current of air removes them from the list of appliances applicable to ordinary conditions, the remarkable results which they afford, not less than their originality, demand for them at least a passing notice. The production of light by the agency of these burners is brought about in a manner altogether different, and is due to quite other causes than those which are concerned in the production of an ordinary illuminating gas-flame. In the latter case, the illuminating power developed is solely due to the hydro-carbons contained in the gas, which are decomposed by the heat of the flame, the separated carbon being raised to a white heat. In the former, the illuminating power is not obtained directly from the gas; but advantage is taken of the heat of the flame, enhanced by the application of a blast of air, to raise to incandescence some refractory foreign material, which latter is thus made to give out light. In the Clamond burner this refractory substance is a basket composed of magnesia, spun into threads; in the Lewis burner it is a cage of platinum wire.

To the unthinking reader it may perhaps appear somewhat surprising that results so remarkable as are yielded by these burners should be obtained, while disregarding, as a source of light, the hydrocarbons contained in gas, and employing them, in common with the other constituents, solely as a source of heat. An explanation, however, is readily forthcoming. As was shown in a former part of this treatise,* the great bulk of ordinary coal gas consists of constituents which, in the act of combustion, produce considerable heat, but scarcely any light; the illuminating power developed in an ordinary gas-flame being almost wholly dependent upon the very small proportion of heavy hydrocarbons which the gas con-

* See Chap. II., p. 21.

F

tains. Thus, the quantity of heat-producing elements contained in the gas being quite disproportionate to the light-yielding hydrocarbons, there is always produced, in an ordinary gas-flame, more heat than is necessary for effectively consuming the free carbon, which is liberated in the flame by the decomposition of the heavy hydrocarbons. This is shown by the fact that coal gas can usually be naphthalized—that is, impregnated with the vapour of naphtha —to a considerable extent before the limit of effective combustion is reached. The object aimed at in the incandescent burners about to be described is to utilize, in the development of illuminating power, the combined heat produced by the combustion of all the constituents of the gas. To this end the heat of combustion is brought to bear upon, and caused to raise to incandescence, some refractory material, extraneous to, but brought within the operation of the flame.

Effect of injecting a blast of air into a gas-flame.

A further explanation of the superior results yielded by these burners may be found in the employment of an artificial blast or current of air. Indeed, without some such arrangement the desired end could not be attained. The heat developed by the unaided flame is diffused over too wide an area to raise the temperature of the heated substance to the necessary degree of incandescence to enable it to give out sufficient light. By injecting a current of air into its midst, the flame is condensed into a smaller compass; and is brought to bear more directly upon the precise locality where its heat may be most effectively employed. Thus, although the total quantity of heat developed remains exactly the same as before, it is concentrated upon a smaller surface of the refractory substance; and the latter is consequently more intensely heated, or, in other words, raised to a more exalted temperature. The very superior illuminating power which is thereby obtained is due to the circumstance that the quantity of light yielded by an incandescent body increases in a higher ratio than the temperature to which it is raised.

Lewis's incandescent gas-burner.

Proceeding now to describe the burners. The one invented by Mr. Lewis (various forms of which are illustrated on the next page) consists of an upright tube, connected at its base to the gas supply, and surmounted by a cap or cage of platinum wire gauze; which latter constitutes a combustion chamber, as it is there that the mixture of gas and air is consumed. Into the lower part of the upright tube the nozzle of an air-pipe is inserted, through which a supply of air can be injected, under pressure, into the burner, after the manner of a blowpipe. There are also small branch tubes leading into the upright gas-tube, and open to the atmosphere. Through these an additional quantity of air enters the burner; being drawn or sucked in by the agency of the main

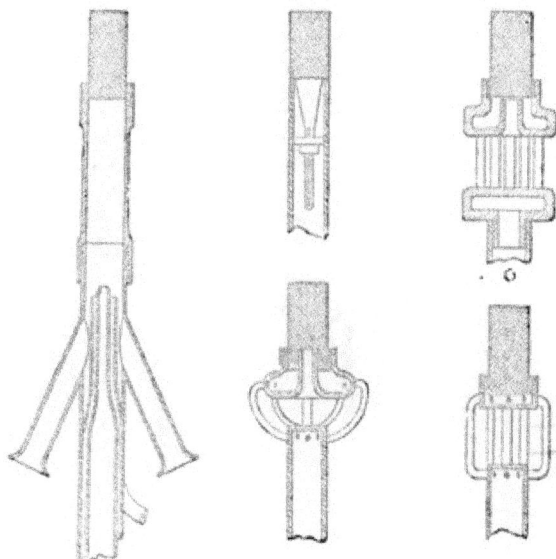

FIG. 26.—LEWIS'S INCANDESCENT GAS-BURNER.

current, which flows through the upright tube. The resemblance to an ordinary Bunsen burner is, therefore, very close. The mixture of gas and air thus produced, when ignited, burns at the platinum cap; the heat which is developed causing the latter to become highly incandescent, and so to give out a brilliant light. To prevent the conduction of heat from the incandescent platinum, through the upright tube, a non-conducting material—such, for instance, as steatite or porcelain—is interposed between the gauze cap and the metal tube.

The light produced by this burner is said to approximate more closely to daylight than that yielded by an ordinary gas-flame (the colours of textile fabrics, for instance, being shown as well by its aid as by daylight); while, on account of its resulting from the incandescence of a fixed body, instead of being emitted from a flame, it is unaffected by a gust of wind, and maintains perfect steadiness under every condition of weather. The illuminating power developed is stated to be equal to 5 standard candles per cubic foot of gas consumed.

M. Clamond's burner, which is shown in fig. 27, is a much more complicated apparatus than the preceding one, and not so easily described; but its main features may be briefly enumerated as follows:—The air (which, as in Mr. Lewis's burner, is supplied under pressure) is divided, as it enters the apparatus, into two portions. One portion is at once mixed with the gas; the remainder being

Clamond's incandescent gas-burner.

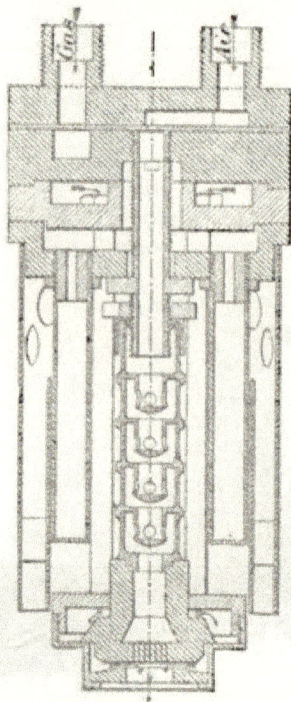

FIG. 27.—CLAMOND'S INCANDESCENT GAS-BURNER.

conveyed, through a peculiarly constructed tube composed of small pieces of refractory material, to the combustion chamber, or "wick," as it is termed, of the burner. This "wick" is a small conical basket, made of a kind of lacework of spun magnesia, which, when raised to incandescence by the heat produced by the combustion of the gas, furnishes the desired illumination. The mixture of gas and air is subdivided, by a "distributor," into two portions, one of which goes direct to the magnesia "wick," there to be burnt, while the other is distributed among a number of tubes, forming so-called "auxiliary burners," the flames of which are utilized to heat the chief air supply; being directed upon the sides of the before-mentioned tube of refractory material, through which it is conveyed. By this means the air is raised to a very high temperature (1000° C., or 1800° Fahr., it is said) before it impinges upon the flame. The result is the production of a most intense heat within the magnesia basket; the latter being raised to brilliant incandescence, and so developing a high illuminating power.

The magnesia basket must be renewed after being in use a period of from 40 to 60 hours, as it gradually deteriorates by the action of the intense heat to which it is subjected; but as the cost is said to be insignificant, this should not be a great drawback. The basket is placed at the base of the burner, in order to obviate the shadow which would otherwise be cast by the apparatus; and it is attached to the main body of the apparatus by platinum wires. As to illuminating power, the only particulars which have been made public refer to the first two models constructed; one of which was said to develop a light equal to that from 6·208 candles, and the other to 9·72 candles per cubic foot of gas consumed.

In a recently designed modification of the burner (which is shown in the accompanying illustration) M. Clamond dispenses

Clamond's new burner.

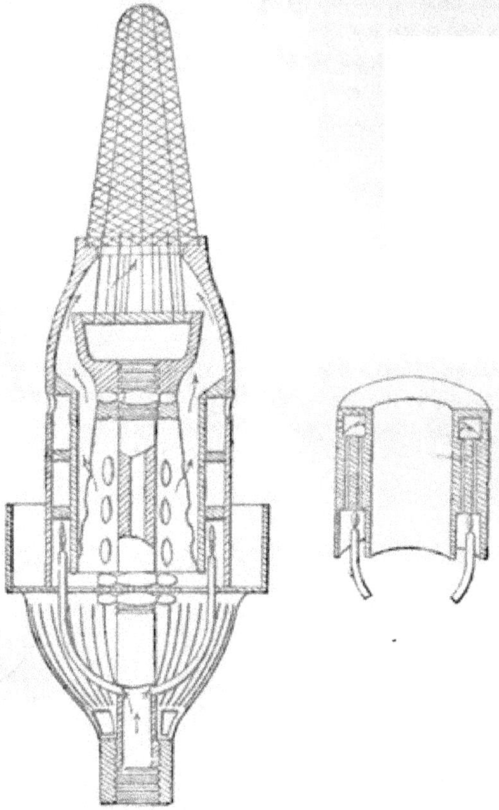

FIG. 28.—CLAMOND'S IMPROVED INCANDESCENT BURNER.

with an artificial supply of air under pressure, and endeavours to obtain similar results by other and simpler means. To this end the position of the magnesia "wick" is reversed (it being

placed at the top of the apparatus); the current of gas is allowed
to draw in upon itself a quantity of air by a precisely similar
arrangement to that adopted in the Bunsen burner; while an
additional supply of air is drawn upon the flame by the accelerated
draught produced by the aid of a glass chimney. As in the more
complicated and complete burner, the air supply is heated by
means of auxiliary burners in the interior of the apparatus. It
has been stated, on the authority of M. Clamond, that this modified burner develops, from the gas consumed, a duty of about
6 candles per cubic foot; being equal to the results yielded by the
more complicated apparatus. Should this be borne out in practice,
M. Clamond will have achieved a noteworthy success. It is, however, advisable to reserve expressing any definite opinion of its
merits until further information is received, or until the burner
has been tried in this country.

CHAPTER VII.

CONCLUSION.

The burners last mentioned may be said to mark the extent of the progress that has been made, down to the present time, in the construction of apparatus for developing light from coal gas; and they remind me that I have arrived at the conclusion of my subject. From the unpretending gas-jet described by Accum— burning, with wonder-provoking steadiness and constancy, " so long as the supply of gas continued "—to the complicated apparatus of M. Clamond, is a long stretch of invention; embracing the labours of many distinct and original workers in the same field, and including numerous variations in the details of burners that have not been touched upon in the foregoing remarks. As was announced in the introduction, I have dealt in this treatise only with the more important or the more successful of the modifications that have been made from time to time in the construction of the gas-burner. In addition to the burners that have been referred to, there have been invented many others, which could not be adequately noticed without prolonging the treatise to an undue length. Some of these (the fruit of much thought and careful experiment) have obtained, in the commercial success that has attended them, no more than their merited reward; others (devoid of any real merit, and in their construction disregarding the most elementary principles of economic combustion) have been brought into somewhat extensive use by the misleading statements and false representations of their inventors, and are only tolerated through the ignorance of the public ; while not a few of the latter class of burners have speedily found the oblivion which they richly deserved. Sufficient, however, has been said to show that many real improvements have been effected in the construction of gas-burners, and to prove that, with the apparatus now available, a far higher duty may be obtained from the gas consumed than was possible only a few years ago.

But although the great advance that has been made in the construction of gas-burners is undoubted, the benefits which ought to result therefrom have not been realized by the gas-consuming public; nor are they likely to be to their full extent. While the ingenious and effective inventions for utilizing the waste heat of combustion, and for lighting by incandescence, may, and doubtless will, in the course of a few years, be far more extensively adopted than at present, it is hardly to be expected that they will be generally employed. Two causes operate to preclude the latter result—namely, their first cost, and the care and attention demanded in their employment. It seems tolerably certain that for a long time yet the great bulk of coal gas, used for lighting purposes, will be consumed through the simple flat-flame burners that have done so much hitherto for the furtherance of gas lighting. Fortunately so much has been done towards the perfection of this class of burners, that, for a very slight expenditure, results may now be obtained far in advance of what could formerly be produced only by the most costly and delicate apparatus. For ordinary situations and requirements, the improved flat-flame burners produced by Bray, Brönner, and Sugg, when intelligently employed, leave scarcely anything to be desired. *When intelligently employed*, I repeat, and with cautious emphasis; for the best of burners will be extravagant and ineffective if employed without due regard to the conditions for which it was made. That which is most needed at the present day, and which will best ensure the continued use of coal gas for the purposes of illumination, is the more general diffusion amongst gas consumers of a knowledge of the principles of combustion, and of the simple precautions to be taken and conditions to be fulfilled in the employment of gas-burners. The apparatus that is available is both varied and effective; what is wanted is the knowledge to use it aright. By contributing to the freer dissemination of that knowledge, purveyors of gas will confer no inconsiderable benefits upon their customers, and, at the same time, will assuredly promote their own interests.

www.ingramcontent.com/pod-product-compliance
Lightning Source LLC
Chambersburg PA
CBHW020330090426
42735CB00009B/1476